INTRODUCTION TO MODERN OPTICS

by Grant R. Fowles

Professor of Physics
University of Utah

Second Edition

DOVER PUBLICATIONS, INC., NEW YORK

This Dover edition, first published in 1989, is an unabridged, corrected republication of the second edition (1975) of the work originally published by Holt, Rinehart and Winston, Inc., New York, 1968.

Library of Congress Cataloging-in-Publication Data

Fowles, Grant R.
  Introduction to modern optics.

  Reprint. Originally published: 2nd ed. New York : Holt, Rinehart, and Winston, 1975.
  Bibliography: p.
  Includes index.
  1. Optics, Physical. I. Title.
QC395.2.F68 1989   535'.2   88-33441
ISBN-13: 978-0-486-65957-2
ISBN-10: 0-486-65957-7

Manufactured in the United States by LSC Communications
65957721   2020
www.doverpublications.com

Preface

Although optics is an old science, in recent years there has been a remarkable upsurge in the importance of optics in both pure science and in technology. This was brought about, in part, by the development of the laser with its rapidly growing list of applications. The obvious need for a modernized undergraduate-level textbook has been the primary reason for writing this book.

This second edition incorporates a number of changes, some minor and some substantial. Part of the text material has been rearranged, and much new material has been added. This includes new problems, expansion of explanatory matter, and updating of certain sections, particularly those dealing with lasers. The sections on relativistic optics, formerly part of the first chapter, has been reordered in the form of an appendix.

The first half of the book deals with classical physical optics: the propagation and polarization of light, coherence and interference, diffraction, and the optical properties of matter. Most of the remainder of the book is devoted to the quantum nature of light: thermal radiation, absorption and emission of light by atoms and molecules, and the theory of optical amplification, and lasers. Also, many applications of the laser to the study of optics are integrated throughout the regular text material.

Chapter 1 treats the propagation of light waves and includes the concepts of phase and group velocities. The vectorial nature of light is taken up in Chapter 2 which also includes the use of the Jones calculus in the study of polarization. Chapter 3 introduces the concepts of partial coherence and coherence length to the study of interference, and includes a brief discussion of the Fourier transform as applied to optics.

Chapter 4 (which was part of Chapter 3 in the first edition) presents a study of multiple-beam interference and includes Fabry-Perot interferometry and multilayer-film theory. Chapter 5 comprises the study of diffraction and includes holography as an application of the theory.

Chapter 6 treats the propagation of light in material media and includes crystal optics and a section on nonlinear optics, a subject which was virtually unheard of until the advent of the laser.

In order to do justice to the theory of light amplification and lasers, treated in Chapter 9, Chapters 7 and 8 offer a brief introduction to the quantum theory of light and elementary optical spectra. These two chapters may be omitted in a short course if the student has already had a course in atomic physics.

The last chapter, Chapter 10, is a brief outline of ray optics and is intended to introduce the student to the matrix method for treating optical systems. The main reason for including this chapter is to apply the ray matrix to the study of laser resonators. A thorough discussion of ray optics is not intended.

The level of the book is such that the student is assumed to have been introduced to Maxwell's equations in an intermediate course in electricity and magnetism. It is further assumed that the student has taken some advanced mathematics beyond calculus so that he is acquainted with such things as elementary matrix algebra, Fourier transforms, and so forth. The level of mathematical proficiency is exemplified by such texts as Wylie's *Advanced Engineering Mathematics*.

For classroom use, a list of problems is included at the end of each chapter. Answers to selected problems are given at the end of the book. The answers to the other problems will be available to teachers on request.

The author wishes to thank all who helped in the preparation of the book. This includes those who used the first edition and submitted many constructive criticisms. Thanks are also due the editorial staff of the publisher and W. E. Wu for his help in proofreading the manuscript.

January, 1975   Grant R. Fowles

Contents

CHAPTER 1

The Propagation
of Light

1.1 Elementary Optical Phenomena and the Nature of Light

"Rays of light," wrote Isaac Newton in his *Treatise on Opticks,* "are very small bodies emitted from shining substances." Newton probably chose to regard light as corpuscular chiefly because of the fact that, in a given uniform medium, light appears to travel in straight-line paths. This is the so-called law of *rectilinear propagation.* The formation of shadows is a familiar example cited to illustrate it.

A contemporary of Newton's, Christiaan Huygens (1629–1695), supported a different description, namely, that light is a "wave motion" spreading out from the source in all directions. The reader will recall the time-honored use of Huygens' construction with primary waves and secondary wavelets to explain the elementary laws of reflection and refraction. Other optical facts that are well explained by the wave picture are interference phenomena such as the formation of bright and dark bands by reflection of light from thin films, and diffraction, or the spreading of light around obstacles.

Owing mainly to the genius of James Clerk Maxwell (1831–1879), we know today that visible light is merely one form of electromagnetic energy, usually described as *electromagnetic waves,* the complete spectrum of which includes radio waves, infrared radiation, the visible spectrum of colors red through violet, ultraviolet radiation, x-rays, and gamma radiation. Furthermore, from the quantum theory of light pioneered by Planck, Einstein, and Bohr during the first two decades of the twentieth century, we know that electromagnetic energy is *quantized;* that is, it can only be imparted to or taken from the electromagnetic field in discrete amounts called *photons.*

Thus the modern concept of light contains elements of both Newton's and Huygens' descriptions. Light is said to have a *dual* nature. Certain phenomena, such as interference, exhibit the wave character of light. Other phenomena, the photoelectric effect, for example, display the particle aspect of light.

2

If one were to ask the question "What is light, really?" there can be no simple answer. There is no familiar object or macroscopic model to employ as an analogy. But understanding need not be based on analogy. A consistent and unambiguous theoretical explanation of all optical phenomena is furnished jointly by Maxwell's electromagnetic theory and the quantum theory. Maxwell's theory treats the *propagation* of light, whereas the quantum theory describes the *interaction* of light and matter or the absorption and emission of light. The combined theory is known as *quantum electrodynamics*. Since electromagnetic theory and quantum theory also explain many other physical phenomena in addition to those related to electromagnetic radiation, it can be fairly assumed that the nature of light is well understood, at least within the context of a mathematical framework that accurately accounts for present experimental observations. The question as to the "true" or "ultimate" nature of light, although as yet unanswered, is quite irrelevant to our study of optics.

1.2 Electrical Constants and the Speed of Light

At a point in empty space the electromagnetic state of the vacuum is said to be specified by two vectors, the electric field $\mathbf{E}$ and the magnetic field $\mathbf{H}$. In the static case, that is, when the two fields do not change with time, $\mathbf{E}$ and $\mathbf{H}$ are independent of one another and are determined, respectively, by the distribution of charges and currents in all space. In the dynamic case, however, the fields are not independent. Their space and time derivatives are interrelated in a manner expressed by curl equations

$$\nabla \times \mathbf{E} = - \mu_0 \frac{\partial \mathbf{H}}{\partial t} \qquad (1.1)$$

$$\nabla \times \mathbf{H} = \epsilon_0 \frac{\partial \mathbf{E}}{\partial t} \qquad (1.2)$$

The divergence conditions

$$\nabla \cdot \mathbf{E} = 0 \qquad (1.3)$$

$$\nabla \cdot \mathbf{H} = 0 \qquad (1.4)$$

indicate the absence of any charge at the point in question. They are true in either the static or the dynamic case.

The four equations above are generally referred to as Maxwell's equations for the vacuum. They can be regarded as the fundamental differential equations of the electromagnetic field in the absence of matter.

The constant μ_0 is known as the *permeability of the vacuum*. It has, by definition, the exact value $4\pi \times 10^{-7}$ henries per meter (H/m).[1] The constant ϵ_0 is called the *permittivity of the vacuum*. Its value must be determined by measurement. The value of ϵ_0 to four significant figures is 8.854×10^{-12} farads per meter (F/m).

Now the fields E and H in the two curl equations can be separated. This is done by taking the curl of one equation and the time derivative of the other and using the fact that the order of differentiation with respect to time or space can be reversed. The result is

$$\nabla \times (\nabla \times E) = -\mu_0\epsilon_0 \frac{\partial^2 E}{\partial t^2} \tag{1.5}$$

$$\nabla \times (\nabla \times H) = -\mu_0\epsilon_0 \frac{\partial^2 H}{\partial t^2} \tag{1.6}$$

Further, by using the divergence conditions (1.3) and (1.4) together with the vector identity

$$\nabla \times (\nabla \times \quad) \equiv \nabla (\nabla \cdot \quad) - \nabla^2(\quad) \tag{1.7}$$

we obtain

$$\nabla^2 E = \frac{1}{c^2} \frac{\partial^2 E}{\partial t^2} \qquad \nabla^2 H = \frac{1}{c^2} \frac{\partial^2 H}{\partial t^2} \tag{1.8}$$

where

$$c = (\mu_0\epsilon_0)^{-1/2} \tag{1.9}$$

Thus the fields satisfy the same formal partial differential equation:

$$\nabla^2(\quad) = \frac{1}{c^2} \frac{\partial^2(\quad)}{\partial t^2}$$

This is called the *wave equation*. It occurs in connection with many different kinds of physical phenomena such as mechanical oscillations of a string, sound waves, vibrating membranes, and so forth [29].[2] The implication here is that *changes* in the fields E and H propagate through empty space with a speed equal to the value of the constant c. In MKS units, c is numerically equal to

$$1/\sqrt{4\pi \times 10^{-7}\epsilon_0} \approx 3 \times 10^8 \text{ meters/second (m/s)}$$

One of the most precise electrical determinations of the value of $(\mu_0\epsilon_0)^{-1/2}$ was made at the National Bureau of Standards by Rosa and Dorsey [33]. They obtained by calculation the capacitance of a condenser of accurately known physical dimensions. This gave the capa-

---

[1] MKS rationalized units are generally used throughout the book. We have chosen to use H rather than B in all equations involving magnetic fields. Wherever it occurs, H can always be written B/μ_0 since nonmagnetic media only are considered in this text.

[2] Numbered references are listed at the end of the book.

citance in electrostatic units. They then measured by means of a bridge the capacitance of the same condenser in electromagnetic units. The ratio of the two values of the capacitance, when converted to MKS units, gives the value of $(\mu_0\epsilon_0)^{-1/2}$. The Rosa and Dorsey result was $2.99784 \cdot 10^8$ m/s to a precision of about one part in 30,000. Other observers had previously measured $(\mu_0\epsilon_0)^{-1/2}$ by purely electrical methods with similar but less precise results.

On the other hand direct measurements of the speed of propagation of light have been carried out by many observers over the years since Römer's historic determination of the speed of light from eclipses of Jupiter's moons in 1676. A summary of measurements of the speed of electromagnetic radiation is given in Table 1.1. The results of all such measurements have always been the same when due account of experimental errors and reduction to vacuum has been made. The conclusion that light is an electromagnetic disturbance is inescapable.

The most precise determination of c has been accomplished by use of the laser. The measurements carried out by K. M. Evanson and co-workers at the National Bureau of Standards in 1972 gave the result

$$c = 299,792,456.2 \pm 1.1 \text{ m/s} \tag{1.10}$$

An excellent review is given in the article "Velocity of Light" by Bergstrand in *The Encyclopedia of Physics* [2].

Speed of Light in a Medium The Maxwell curl equations for the electric and magnetic fields in isotropic nonconducting media are precisely the same as those for the vacuum, except that the vacuum constants μ_0 and ϵ_0 are replaced by the corresponding constants for the medium, namely, μ and ϵ. Consequently, the speed of propagation u of electromagnetic fields in a medium is given by

$$u = (\mu\epsilon)^{-1/2} \tag{1.11}$$

By introducing the two dimensionless ratios

$$K = \frac{\epsilon}{\epsilon_0} \tag{1.12}$$

called the relative permittivity or the dielectric constant, and

$$K_m = \frac{\mu}{\mu_0} \tag{1.13}$$

known as the relative permeability, we can write

$$u = (\mu\epsilon)^{-1/2} = (K_m\mu_0 K\epsilon_0)^{-1/2} = c(KK_m)^{-1/2} \tag{1.14}$$

The *index of refraction n* is defined as the ratio of the speed of

Table 1.1.   MEASUREMENTS OF THE SPEED OF ELECTROMAGNETIC
RADIATION

——————— A. SPEED OF LIGHT ———————

Date	Investigator	Method	Result (km/s)	
1849	Fizeau	Rotating toothed wheel	313,000	± 5000*
1850	Foucault	Rotating mirror	298,000	± 2000*
1875	Cornu	Rotating mirror	299,990	± 200
1880	Michelson	Rotating mirror	299,910	± 150
1883	Newcomb	Rotating mirror	299,860	± 30
1928	Mittelstaedt	Kerr cell shutter	299,778	± 10
1932	Pease and Pearson	Rotating mirror	299,774	± 2
1940	Hüttel	Kerr cell shutter	299,768	± 10
1941	Anderson	Kerr cell shutter	299,776	± 6
1951	Bergstrand	Kerr cell shutter	299,793.1	± 0.3

——————— B. SPEED OF RADIO WAVES ———————

Date	Investigator	Method	Result (km/s)	
1923	Mercier	Standing waves on wires	299,782	± 30
1947	Jones and Conford	Oboe radar	299,782	± 25
1950	Bol	Cavity resonator	299,789.3	± 0.4
1950	Essen	Cavity resonator	299,792.5	± 3.0
1951	Aslakson	Shoran radar	299,794.2	± 1.9
1952	Froome	Microwave Interferometer	299,792.6	± 0.7

——————— C. RATIO OF ELECTRICAL UNITS ———————

Date	Investigator	Result	
1857	Weber and Kohlrausch	310,000	± 20,000*
1868	Maxwell	288,000	± 20,000*
1883	Thomson	282,000	± 20,000*
1907	Rosa and Dorsey	299,784	± 10

* Estimated error limits.

light in vacuum to its speed in the medium. Hence

$$\frac{c}{u} = n = (KK_m)^{1/2} \tag{1.15}$$

Most transparent optical media are nonmagnetic so that $K_m = 1$, in which case the index of refraction should be equal to the square root of the relative permittivity:

$$n = \sqrt{K} \qquad (1.16)$$

Table 1.2 shows a number of examples in which the index of refraction is compared to the square root of the static permittivity.

Table 1.2. INDEX OF REFRACTION VERSUS THE SQUARE ROOT OF THE STATIC PERMITTIVITY [14]

Substance	n (Yellow Light)	$\sqrt{K}$
Air (1 atm)	1.0002926	1.000295
CO_2 (1 atm)	1.00045	1.0005
Polystyrene	1.59	1.60
Glass*	1.5–1.7	2.0–3.0
Fused quartz	1.46	1.94
Water	1.33	9.0
Ethyl alcohol	1.36	5.0

* Approximate values.

The agreement is good in the case of the gases, air, and carbon dioxide, and also for nonpolar solids, such as polystyrene. For media that contain polar molecules, such as water and alcohol, the agreement is poor. This is due to the high static polarizability of these substances.

Actually, the index of refraction is found to vary with the frequency of the radiation. This is true for all transparent optical media. The variation of the index of refraction with frequency is called *dispersion*. The dispersion of glass is responsible for the familiar splitting of light into its component colors by a prism.

In order to explain dispersion it is necessary to take into account the actual motion of the electrons in the optical medium through which the light is traveling. The theory of dispersion will be treated in detail later in Chapter 6.

1.3 Plane Harmonic Waves. Phase Velocity

If we employ rectangular coordinates and resolve the vector wave equations (1.8) and (1.9) into components, we observe that each component of **E** and **H** satisfies the general scalar[3] wave equation

$$\frac{\partial^2 U}{\partial x^2} + \frac{\partial^2 U}{\partial y^2} + \frac{\partial^2 U}{\partial z^2} = \frac{1}{u^2} \frac{\partial^2 U}{\partial t^2} \qquad (1.17)$$

[3] The vectorial nature of electromagnetic waves is treated in Chapter 2.

Here the quantity U stands for any one of the field components E_x, E_y, E_z, H_x, H_y, H_z.

Waves in One Dimension   Now, for the moment, let us consider the special case in which the spatial variation of U occurs only in some particular coordinate direction, say the z direction. In this case the operator ∇^2 reduces to $\partial^2/\partial z^2$, and Equation (1.17) becomes the one-dimensional wave equation

$$\frac{\partial^2 U}{\partial z^2} = \frac{1}{u^2}\frac{\partial^2 U}{\partial t^2} \tag{1.18}$$

By direct substitution it is easy to verify that the function

$$U(z,t) = U_0 \cos (kz - \omega t) \tag{1.19}$$

is, in fact, a solution of our wave equation (1.18) provided that the ratio of the constants ω and k is equal to the constant u, namely,

$$\frac{\omega}{k} = u \tag{1.20}$$

The particular solution given by Equation (1.19) is fundamental to the study of optics. It represents what is known as a *plane harmonic wave*. A graph is shown in Figure 1.1. At a given instant in time the

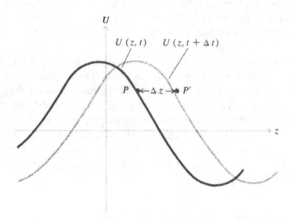

Figure 1.1.   Graph of U versus z at times t and $t + \Delta t$.

wave function $U(z,t)$ varies sinusoidally with distance z, and at a given fixed value of z the wave function varies harmonically with time. The progressive nature of the wave is illustrated by drawing two curves, $U(z,t)$ and $U(z,t + \Delta t)$. The latter curve is displaced in the positive z direction by a distance

$$\Delta z = u \, \Delta t$$

This is the distance between any two points of corresponding phase, say PP', as indicated in the figure. This is the reason that u is called the *phase velocity*. Clearly the function $U_0 \cos (kz + \omega t)$ represents a wave moving in the *negative* z direction.

The constants ω and k are called the *angular frequency* and *angular wavenumber*, respectively.[4] The *wavelength* λ is defined as the distance, measured along the direction of propagation, such that the wave function goes through one complete cycle. The reciprocal of the wavelength is known as the *spectroscopic wavenumber* and is designated by the symbol σ. The time for one complete cycle is called the *period*, denoted by T. The number of cycles per unit of time is called the *frequency* ν. According to the above definitions, the wave will travel a distance λ in time T. The following relationships among the various parameters are easily verified:

$$\lambda = uT = \frac{2\pi}{k} = \frac{1}{\sigma} \tag{1.21}$$

$$\nu = \frac{u}{\lambda} = \frac{\omega}{2\pi} = \frac{1}{T} \tag{1.22}$$

Waves in Three Dimensions   Returning now to the three-dimensional wave equation (1.17), it is readily verified that this equation is satisfied by a three-dimensional plane harmonic wave function as follows:

$$U(x,y,z,t) = U_0 \cos (\mathbf{k} \cdot \mathbf{r} - \omega t) \tag{1.23}$$

where the position vector $\mathbf{r}$ is defined as

$$\mathbf{r} = \hat{\mathbf{i}}x + \hat{\mathbf{j}}y + \hat{\mathbf{k}}z$$

and the *propagation vector* or *wave vector* $\mathbf{k}$ is given in terms of its components by

$$\mathbf{k} = \hat{\mathbf{i}}k_x + \hat{\mathbf{j}}k_y + \hat{\mathbf{k}}k_z \tag{1.24}$$

The magnitude of the wave vector is equal to the wavenumber, which was previously defined; that is,

$$|\mathbf{k}| = k = (k_x^2 + k_y^2 + k_z^2)^{1/2} \tag{1.25}$$

In order to interpret Equation (1.23), consider the argument of the cosine, namely, $\mathbf{k} \cdot \mathbf{r} - \omega t$. Constant values of this quantity define a

---

[4] Some writers prefer to call them merely *frequency* and *wavenumber*. The constant k is also called the *propagation constant*.

set of planes in space called *surfaces of constant phase,*

$$\mathbf{k} \cdot \mathbf{r} - \omega t = k_x x + k_y y + k_z z - \omega t = constant \qquad (1.26)$$

It follows that the direction cosines of the planes of constant phase are proportional to the components of the propagation vector **k**. This means that **k** is normal to the wave surfaces, as shown in Figure 1.2. Furthermore, by virtue of the time factor in Equation (1.26) we see that these wave surfaces move in the direction of **k** at a rate equal to the phase velocity. Explicitly,

$$u = \frac{\omega}{k} = \frac{\omega}{\sqrt{k_x^2 + k_y^2 + k_z^2}} \qquad (1.27)$$

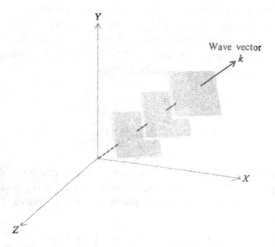

Figure 1.2. Equiphase surfaces in a plane wave.

Sources of Electromagnetic Waves   Electromagnetic radiation is created by oscillating electric charges. The frequency of oscillation determines the kind of radiation that is emitted. The various portions of the electromagnetic spectrum, designated according to frequency and wavelength, are shown in Table 1.3. (The quantum energy of the radiation is also listed. Quantum theory is discussed later in Chapter 7.)

The units of wavelength commonly used for the optical region are the following:

UNIT	ABBREVIATION	EQUIVALENT
micron	μ	10^{-6} m
nanometer	nm	10^{-9} m
angstrom	Å	10^{-10} m

Table 1.3. The Electromagnetic Spectrum

Type of Radiation		Frequency	Wavelength	Quantum Energy
"Wave" region	radio waves	10^9 Hz and less	300 mm and longer	0.000004 eV and less
	microwaves	10^9 Hz to 10^{12} Hz	300 mm to 0.3 mm	0.000004 eV to 0.004 eV
"Optical" region	infrared	10^{12} Hz to 4.3×10^{14} Hz	$300~\mu$ to $0.7~\mu$	0.004 eV to 1.7 eV
	visible	4.3×10^{14} Hz to 5.7×10^{14} Hz	$0.7~\mu$ to $0.4~\mu$	1.7 eV to 2.3 eV
	ultraviolet	5.7×10^{14} Hz to 10^{16} Hz	$0.4~\mu$ to $0.03~\mu$	2.3 eV to 40 eV
"Ray" region	x-rays	10^{16} Hz to 10^{19} Hz	300 Å to 0.3 Å	40 eV to 40,000 eV
	gamma rays	10^{19} Hz and above	0.3 Å and shorter	40,000 eV and above

Note: The numerical values are only approximate and the division into the various regions are for illustration only. They are quite arbitrary.

The unit of frequency is the cycle per second also called the *hertz* (Hz).

If in a given source the charges all oscillate in unison, the source is said to be *coherent*. If the charges oscillate independently and randomly, the source is called *incoherent*. Ordinary sources of radiation in the optical region are incoherent, for example, tungsten filament lamps, fluorescent lamps, flames, and so forth.

Man-made sources of radio waves and microwaves are normally coherent. These low-frequency coherent sources are basically electronic oscillators that utilize amplifying devices such as vacuum tubes, transistors, klystrons, and so forth. The development of optical amplification, the laser, has extended the range of coherent sources to the optical region of the electromagnetic spectrum. The theory of the laser is treated in Chapter 9.

1.4 Alternative Ways of Representing Harmonic Waves

Let the unit vector $\hat{n}$ denote the direction of the wave vector $\mathbf{k}$. Then $\mathbf{k} = \hat{n}k$ and, accordingly, the expression for a plane harmonic wave Equation (1.23) is equivalent to

$$U_0 \cos \left[(\hat{n} \cdot \mathbf{r} - ut)k \right]$$

It should be noted that the order of writing the factors in the argument of the cosine is immaterial, since $\cos \theta = \cos (-\theta)$. It is also immaterial whether one uses a cosine function or a sine function. Both represent the same thing, except for phase.

It should also be noted here that the wave function can be represented in various ways by use of the Equations (1.21) and (1.22) as found earlier (see Problem 1.1).

The Complex Wave Function It is often convenient to make use of the identity

$$e^{i\theta} = \cos \theta + i \sin \theta$$

and write

$$U = U_0 e^{i(\mathbf{k} \cdot \mathbf{r} - \omega t)} \tag{1.28}$$

to represent a plane harmonic wave. It is understood that the real part is the actual physical quantity being represented. The real part is identical with the previous expression (1.23). However, it is easy to verify that the complex expression (1.28) is itself a solution of the wave equation. The main reason for using the complex exponential expression is that it is algebraically simpler than the trigonometric expression. An example of the use of the complex exponential is given in the following section.

Spherical Waves The functions $\cos(kr - \omega t)$ and $e^{i(kr-\omega t)}$ have constant values on a sphere of any given radius r at a given time t. As t increases, the functions would represent spherical expanding waves except for the fact that they are not solutions of the wave equation. However, it is easy to verify that the functions

$$\frac{1}{r}\cos(kr - \omega t) \qquad \text{and} \qquad \frac{1}{r}e^{i(kr-\omega t)}$$

are, indeed, solutions of the wave equation and therefore represent spherical waves propagating outward from the origin (see Problem 1.2).

1.5 Group Velocity

Suppose we have two harmonic waves that have slightly different angular frequencies. Let us denote these frequencies by $\omega + \Delta\omega$ and $\omega - \Delta\omega$, respectively. The corresponding wavenumbers will, in general, also differ. We shall denote them by $k + \Delta k$ and $k - \Delta k$. Now suppose, in particular, that the two waves have the same amplitude, U_0, and are traveling in the same direction, say the z direction. Then the superposition of the two waves is given, in complex notation, by

$$U = U_0 e^{i[(k+\Delta k)z - (\omega + \Delta\omega)t]} + U_0 e^{i[(k-\Delta k)z - (\omega - \Delta\omega)t]} \qquad (1.29)$$

By factoring and collecting terms, we get

$$U = U_0 e^{i(kz-\omega t)}[e^{i(z\Delta k - t\Delta\omega)} + e^{-i(z\Delta k - t\Delta\omega)}] \qquad (1.30)$$

or

$$U = 2U_0 e^{i(kz-\omega t)}\cos(z\,\Delta k - t\,\Delta\omega) \qquad (1.31)$$

The final expression can be interpreted as a single wave $2U_0 e^{i(kz-\omega t)}$, which has a modulation envelope $\cos(z\,\Delta k - t\,\Delta\omega)$ as shown in Figure 1.3. This modulation envelope does not travel at the phase velocity ω/k of the individual waves, but rather at a rate $\Delta\omega/\Delta k$, called the *group velocity*. We shall denote the group velocity by u_g. Then

$$u_g = \frac{\Delta\omega}{\Delta k} \qquad (1.32)$$

or, in the limit

$$u_g = \frac{d\omega}{dk} \qquad (1.33)$$

Now, in all optical media the phase velocity u is a function of the angular frequency ω. This is the phenomenon of dispersion, mentioned earlier. In a medium in which the index of refraction, $n = c/u$, varies in a known way with frequency or wavelength, we can write

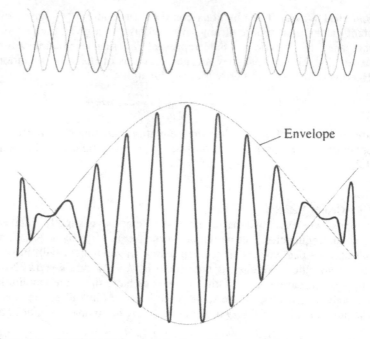

Figure 1.3. Envelope of the combination of two harmonic waves.

$$\omega = ku = \frac{kc}{n} \tag{1.34}$$

Hence

$$u_g = \frac{d\omega}{dk} = \frac{d}{dk}\left(\frac{kc}{n}\right) = \frac{c}{n} - \frac{ck}{n^2}\frac{dn}{dk} = u\left(1 - \frac{k}{n}\frac{dn}{dk}\right) \tag{1.35}$$

For purposes of practical calculation of the group velocity, the following formulas, which are left as problems, are often used:

$$u_g = u - \lambda\frac{du}{d\lambda} \tag{1.36}$$

$$\frac{1}{u_g} = \frac{1}{u} - \frac{\lambda_0}{c}\frac{dn}{d\lambda_0} \tag{1.37}$$

where λ_0 is the vacuum wavelength.

We see that for a medium in which the phase velocity, or the index of refraction, is constant, the phase and group velocities do not differ. In particular, in the vacuum we have

$$u_g = u = c \qquad (1.38)$$

For most optical media the index of refraction increases with increasing frequency, or wavenumber, so that dn/dk is positive. For such media the group velocity is therefore less than the phase velocity. Since any signal can be considered as a modulation of some type imposed on a continuous wave, a signal will travel at the group velocity and thus propagate at a slower speed than the phase velocity, generally. In this case the individual waves within a given modulation envelope will appear at the trailing end of the envelope, travel to the forward edge, and disappear. The opposite would occur in those cases in which the group velocity was greater than the phase velocity.

It is clear that the group velocity itself is, in general, a function of frequency. However, if a given modulated wave occupies a narrow range of frequencies, then the group velocity is well defined and unique. Such would be the case for pulses of nearly monochromatic light. Michelson was one of the first to demonstrate experimentally the difference between phase and group velocity. He found that the speed of light pulses in carbon disulfide was $c/1.76$, for yellow light, whereas the index of refraction is 1.64, so the phase velocity is $c/1.64$.

In any determination of the velocity of light by a time-of-flight method, account must be taken of the difference between the phase velocity and the group velocity in a medium. Appropriate corrections must be made when computing the final result from the experimental data.

1.6 The Doppler Effect

If a source of waves and a receiver are in relative motion while the waves are being received, the observed frequency is changed compared with that in which there is no motion. This well-known phenomenon was first studied in connection with sound waves by J. C. Doppler. An elementary analysis proceeds as follows. If the source is moving away from the receiver with a velocity u, the number ν of waves emitted per second will be expanded into a distance $c + u$ rather than c, where c is the speed of the waves in the medium. Here the medium is considered to be at rest with respect to the receiver. The observed frequency ν', being the number of waves reaching the receiver per second, will then be

$$\nu' = \nu \left(\frac{c}{c + u}\right) = \nu \left(1 - \frac{u}{c} + \frac{u^2}{c^2} - \cdots\right) \qquad (1.39)$$

On the other hand, if the receiver is moving away from the source, the source considered to be stationary in the medium, then the speed of the waves relative to the receiver will be $c - u$, and therefore the observed frequency will be given by

$$\nu' = \nu \left(\frac{c - u}{c} \right) = \nu \left(1 - \frac{u}{c} \right) \qquad (1.40)$$

or

$$\frac{\nu - \nu'}{\nu} = \frac{\Delta \nu}{\nu} = \frac{u}{c} \qquad (1.41)$$

If the source and receiver are moving toward one another, then the sign of u is changed in each of the above formulas.

It can be seen from the series expansion of Equation (1.39) that if the value of u is very small in comparison with the wave velocity c, then the quadratic and higher terms can be neglected. The two cases then give the same result.

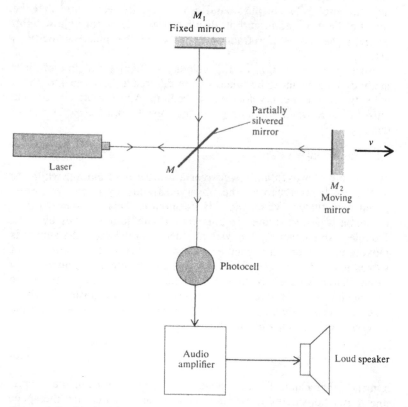

Figure 1.4. Method for observing the Doppler effect with a laser.

Measurable Doppler shifts in the case of light waves[5] can be observed in the laboratory by means of atomic beams, that is, by the light emitted from a beam of high-speed atoms. Another method is to reflect light from a moving mirror. With ordinary light sources it is necessary to have the mirror move at a very high speed, that is, by attaching it to a rapidly rotating wheel. But by using a laser as the source, the Doppler effect can be observed with speeds of only a few centimeters per second. The experimental arrangement is shown in Figure 1.4. The light from the laser is split into two beams by means of a lightly silvered mirror M. One of the beams is reflected by a fixed mirror M_1 back through M to a photocell P. The other beam is reflected by a moving mirror M_2. The two beams combine at P to produce a beat note whose frequency is equal to the difference $\Delta \nu$ between the frequencies of the two beams. If u_m is the speed of the moving mirror, then $\Delta \nu / \nu = 2u_m / c$. The factor 2 arises from the fact that the apparent speed of the virtual source produced by the moving mirror is just twice the speed of the mirror.

Doppler shifts of spectral lines are well known in astronomy. The effect is used to determine the motion of astronomical objects. For example, in the case of binary stars—that is, two stars revolving about their common center of mass—the spectrum lines show a periodic doubling due to the fact that one star approaches the earth and the other recedes from the earth in a regular fashion as illustrated in Figure 1.5.

Typical astronomical velocities are of the order of 100 kilometers per second (km/s) so that u/c is of the order of 10^{-4}. In the case of the very distant galaxies, however, the spectrum lines are shifted to lower frequencies by amounts that indicate recessional velocities up to half the speed of light. This shift, known as the *galactic red shift*, seems to be proportional to the distance and has therefore been interpreted as indicating an expansion of the universe. The recently discovered quasistellar objects, or *quasars*, have very large red shifts, indicating recessional speeds up to $0.8c$.

Relativity Correction to the Doppler Formula   In Appendix I, which treats relativistic optics, it is shown that according to the special theory of relativity there is no distinction between the two cases, "observer in motion" and "source in motion," given by Equations (1.39) and (1.40) above. Rather there is only relative motion, and the relativistic formula is shown to be

$$\nu' = \nu \sqrt{\frac{1 - u/c}{1 + u/c}} = \nu \left(1 - \frac{u}{c} + \frac{1}{2}\frac{u^2}{c^2} - \cdots\right) \qquad (1.42)$$

---

[5] Fizeau was one of the first to study the effect in connection with light waves. For this reason, the Doppler effect in light is also known as the *Doppler-Fizeau effect*.

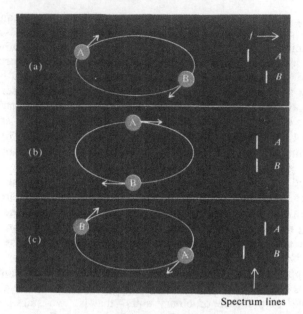

Spectrum lines

Figure 1.5.   Illustration of the motion of a binary star system and the Doppler shifts of a spectrum line.

Here u is positive if the source and receiver are moving away from one another, and negative if they are moving toward each other. The series expansion shows that the relativistic Doppler shift differs from the nonrelativistic shift only in the second order of u/c. The difference is negligible for small velocities, but becomes important when u approaches c.

Doppler Broadening of Spectrum Lines   Another way in which the Doppler effect is manifest is the widening of the spectrum lines from a gaseous discharge. This widening is due to the random thermal motion of the radiating atoms. According to elementary kinetic theory [31], the value of the root mean square of a given component of the velocity of an atom in a gas is equal to $\sqrt{kT/m}$, where T is the absolute temperature, k is Boltzmann's constant, and m is the mass of the atom. At any instant, part of the atoms are going toward an observer and part are going away. The "half-power" width $\Delta\nu$ of a spectrum line of mean frequency ν, due to thermal motion, is given by the relationship

$$\frac{\Delta\nu}{\nu} = \frac{2\sqrt{2\ln 2}}{c} \sqrt{\frac{kT}{m}} \tag{1.43}$$

The numerical factor $2\sqrt{2\ln 2}$ comes from the distribution of the velocities, which is a *Gaussian* function [40]. The intensity distribution, as a function of frequency, is also Gaussian.

We see that the width is proportional to the square root of the temperature and is inversely proportional to the square root of the mass of the atom. Thus hydrogen, the lightest atom, gives the widest spectrum lines at a given temperature. To obtain the narrowest lines, the discharge is cooled and heavy atoms are used. Thus the international standard of length is the wavelength of the orange krypton line emitted by a liquid air-cooled discharge through krypton 86 gas. This provides a precise, highly reproducible source for interferometric measurements. It is likely, however, that the krypton standard of lengths will be replaced by a laser standard in the near future.

PROBLEMS

1.1 Express the wave function for one-dimensional harmonic waves in terms of the following pairs of parameters:
   (a) Frequency and wavelength
   (b) Period and wavelength
   (c) Angular frequency and phase velocity
   (d) Wavelength and phase velocity

1.2 Show that $f(z - ut)$ is a solution of the one-dimensional wave equation $\partial^2 f/\partial z^2 = (1/u^2)\, \partial^2 f/\partial t^2$, where f is any differentiable function of the single argument $z - ut$.

1.3 The three-dimensional generalization of the preceding problem is the function $f(\hat{n} \cdot \mathbf{r} - ut)$, where $\hat{n}$ is a unit vector. Verify that the three-dimensional wave equation $\nabla^2 f = (1/u^2)\, \partial^2 f/\partial t^2$ is satisfied by the function f in this case.

1.4 Prove that the spherical harmonic wave function

$$\frac{1}{r}\, e^{i(kr - \omega t)}$$

is a solution of the three-dimensional wave equation, where $r = (x^2 + y^2 + z^2)^{1/2}$. The proof is easier if spherical coordinates are used.

1.5 A helium–neon laser emits light of wavelength 633 nm (vacuum wavelength). Find the numerical value of the angular wavenumber k of this radiation in water ($n = 1.33$).

1.6 Derive the formulas

$$u_g = u - \lambda\, \frac{du}{d\lambda}$$

and

$$\frac{1}{u_g} = \frac{1}{u} - \frac{\lambda_0}{c}\, \frac{dn}{d\lambda_0}$$

where u_g is the group velocity, u is the phase velocity, λ is the wavelength, λ_0 is the vacuum wavelength, and n is the index of refraction.

1.7 The variation of index of refraction with wavelength, in the case of a transparent substance such as glass, can be represented approximately by an empirical equation of the form (known as *Cauchy's equation*)

$$n = A + B\lambda_0^{-2}$$

in which A and B are empirical constants, and λ_0 is the vacuum wavelength. Find the group velocity at $\lambda_0 = 500$ nm for a particular type of glass for which $A = 1.50$ and $B = 3 \times 10^4$ (nm)^2.

1.8 The index of refraction of a certain hypothetical substance is found to vary inversely with the vacuum wavelength; that is, $n = A/\lambda_0$. Show that the group velocity at a given wavelength is just half the phase velocity.

1.9 Two waves have slightly different frequencies and wavelengths, namely, ν and $\nu + \Delta\nu$, and λ and $\lambda + \Delta\lambda$, respectively. Show that the ratios $|\Delta\nu/\nu|$ and $|\Delta\lambda/\lambda|$ are approximately equal.

1.10 The two members of a binary star system are a distance d apart, and they revolve about their common center of mass with angular velocity ω. Show that the Doppler splitting $\Delta\lambda$ between the spectral lines from the two stars has a maximum value of $\lambda\omega \, d/c$, provided the orbital plane of the star system is edgewise as seen from the earth. What would the splitting be if the orbital plane was inclined at a certain angle θ?

1.11 Calculate the Doppler width of a spectrum line from a neon discharge tube operating at a temperature of 100°C. The value of the Boltzmann constant k is 1.38×10^{-23} joules per degree Kelvin (J/°K) and the mass of the neon atom is 3.34×10^{-26} kg. Take $\lambda = 600$ nm and find the line width in both frequency and wavelength.

CHAPTER 2

The Vectorial
Nature of Light

2.1 General Remarks

The various Cartesian components of the fields in an electromagnetic wave, as we have shown in the previous chapter, individually satisfy the same basic wave equation:

$$\nabla^2 U = \frac{1}{u^2} \frac{\partial^2 U}{\partial t^2} \tag{2.1}$$

The Maxwell curl equations require that for fields that vary in time and space, a magnetic field must always accompany an electric field and vice versa. In particular, for electromagnetic waves there exists a very definite relationship between the two fields.

We wish to examine this relationship in detail. It will be useful at this juncture to establish some operator identities in connection with plane harmonic waves. Consider the complex exponential expression for a plane harmonic wave

$$\exp i(\mathbf{k} \cdot \mathbf{r} - \omega t)$$

Taking the time derivative, we have

$$\frac{\partial}{\partial t} \exp i(\mathbf{k} \cdot \mathbf{r} - \omega t) = -i\omega \exp i(\mathbf{k} \cdot \mathbf{r} - \omega t) \tag{2.2}$$

and taking the partial derivative with respect to one of the space variables, say x, we get

$$\frac{\partial}{\partial x} \exp i(\mathbf{k} \cdot \mathbf{r} - \omega t) = \frac{\partial}{\partial x} \exp i(k_x x + k_y y + k_z z - \omega t)$$

$$= ik_x \exp i(\mathbf{k} \cdot \mathbf{r} - \omega t)$$

Hence on application of the del operator

$$\nabla = \hat{\mathbf{i}} \frac{\partial}{\partial x} + \hat{\mathbf{j}} \frac{\partial}{\partial y} + \hat{\mathbf{k}} \frac{\partial}{\partial z} \tag{2.3}$$

it readily follows that

$$\nabla \exp i(\mathbf{k} \cdot \mathbf{r} - \omega t) = i\mathbf{k} \exp i(\mathbf{k} \cdot \mathbf{r} - \omega t) \tag{2.4}$$

22

Thus we have the following operator relations:

$$\frac{\partial}{\partial t} \rightarrow -i\omega \tag{2.5}$$

$$\nabla \rightarrow i\mathbf{k} \tag{2.6}$$

which are valid for plane harmonic waves. (The reader is reminded here that $\hat{\mathbf{i}}$, $\hat{\mathbf{j}}$, and $\hat{\mathbf{k}}$ are unit coordinate vectors, whereas i is the square root of -1, and $\mathbf{k}$ is the wave vector. This notation may be confusing, but it is standard.)

Let us return now to the Maxwell equations for isotropic nonconducting media:

$$\nabla \times \mathbf{E} = -\mu \frac{\partial \mathbf{H}}{\partial t} \tag{2.7}$$

$$\nabla \times \mathbf{H} = \epsilon \frac{\partial \mathbf{E}}{\partial t} \tag{2.8}$$

$$\nabla \cdot \mathbf{E} = 0 \tag{2.9}$$

$$\nabla \cdot \mathbf{H} = 0 \tag{2.10}$$

From the operator relations, Equations (2.5) and (2.6), the Maxwell equations take the following form for plane harmonic waves:

$$\mathbf{k} \times \mathbf{E} = \mu\omega\mathbf{H} \tag{2.11}$$

$$\mathbf{k} \times \mathbf{H} = -\epsilon\omega\mathbf{E} \tag{2.12}$$

$$\mathbf{k} \cdot \mathbf{E} = 0 \tag{2.13}$$

$$\mathbf{k} \cdot \mathbf{H} = 0 \tag{2.14}$$

A study of the above equations shows that the three vectors $\mathbf{k}$, $\mathbf{E}$, and $\mathbf{H}$ constitute a mutually orthogonal triad. The electric and magnetic fields are perpendicular to one another, and they are both perpendicular to the direction of propagation as illustrated in Figure 2.1. It also follows that the magnitudes of the fields are related according to the equations

$$H = \frac{\epsilon\omega}{k} E = \epsilon u E \tag{2.15}$$

where we have used the fact that the phase velocity $u = \omega/k$. Further, in terms of the index of refraction $n = c/u$ we have

$$H = \frac{nE}{Z_0} \tag{2.16}$$

in which the quantity

$$Z_0 = (\mu_0/\epsilon_0)^{1/2}$$

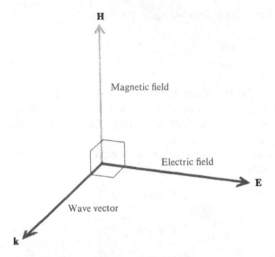

Figure 2.1. Relationships among the field vectors and the wave vector in an electromagnetic wave.

is known as the *impedance of free space*. Numerically its value is about 377 ohms. Equation (2.16) shows that the ratio of the magnetic field to the electric field of an electromagnetic wave propagating through a medium is proportional to the index of refraction of the medium. Thus when a beam of light passes from air into glass, $n = 1.5$, the ratio of the magnetic to electric field suddenly increases by the factor 1.5.

2.2   Energy Flow. The Poynting Vector

Poynting's theorem [16] states that the time rate of flow of electromagnetic energy per unit area is given by the vector $\mathbf{S}$, called the *Poynting vector*, defined as the cross product of the electric and magnetic fields,

$$\mathbf{S} = \mathbf{E} \times \mathbf{H} \qquad (2.17)$$

This vector specifies both the direction and the magnitude of the energy flux. In the MKS system of units $\mathbf{S}$ is expressed in watts per square meter.[1]

Consider now the case of plane harmonic waves in which the fields are given by the real expressions

$$\mathbf{E} = \mathbf{E}_0 \cos (\mathbf{k} \cdot \mathbf{r} - \omega t) \qquad (2.18)$$

$$\mathbf{H} = \mathbf{H}_0 \cos (\mathbf{k} \cdot \mathbf{r} - \omega t) \qquad (2.19)$$

[1] In Gaussian units $\mathbf{S} = (c/4\pi)\, \mathbf{E} \times \mathbf{H}$.

We then have

$$S = E_0 \times H_0 \cos^2 (k \cdot r - \omega t) \tag{2.20}$$

for the instantaneous value of the Poynting vector. Since the average value of the cosine squared is just $\frac{1}{2}$, then for the average value of the Poynting vector, we can write

$$\langle S \rangle = \tfrac{1}{2} E_0 \times H_0 \tag{2.21}$$

(If the complex exponential form of the wave functions for E and H is used, the average Poynting flux can be expressed as $\frac{1}{2} E_0 \times H_0^*$. See Problem 2.4.)

Since the wave vector k is perpendicular to both E and H, it has the same direction as the Poynting vector. Consequently, an alternative expression for the average Poynting flux is

$$\langle S \rangle = I \frac{k}{k} = I \, \hat{n} \tag{2.22}$$

in which $\hat{n}$ is a unit vector in the direction of propagation and I is the magnitude of the average Poynting flux. The quantity I is called the *irradiance*.[2] It is given by

$$I = \frac{1}{2} E_0 H_0 = \frac{n}{2 Z_0} |E_0|^2 \tag{2.23}$$

The last step follows from the relations between the magnitudes of the electric and magnetic vectors developed in the previous section. Thus the rate of flow of energy is proportional to the square of the amplitude of the electric field. In isotropic media the direction of the energy flow is specified by the direction of S and is the same as the direction of the wave vector k. (In nonisotropic media, for example crystals, S and k are not always in the same direction. This will be discussed later in Chapter 6.)

2.3 Linear Polarization

Consider a plane harmonic electromagnetic wave for which the fields E and H are given by the expressions

$$E = E_0 \exp i(k \cdot r - \omega t) \tag{2.24}$$

$$H = H_0 \exp i(k \cdot r - \omega t) \tag{2.25}$$

If the amplitudes E_0 and H_0 are constant real vectors, the wave is said to be *linearly polarized* or plane polarized. We know from the theory

---

[2] Sometimes the word *intensity* is used for I, but this is not technically correct (see Chapter 7).

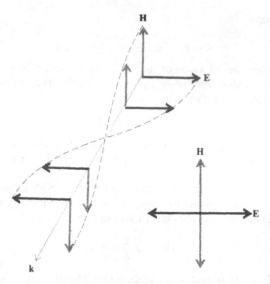

Figure 2.2. Fields in a plane wave, linearly polarized.

of the previous section that the fields **E** and **H** are mutually perpendicular. It is traditional in optics to designate the direction of the electric field as the direction of polarization. Figure 2.2 shows a diagram of the fields in a plane, linearly polarized wave.

In the case of natural, or so-called unpolarized, light the instantaneous polarization fluctuates rapidly in a random manner. A *linear polarizer* is a device that produces linearly polarized light from unpolarized light. There are several kinds of linear polarizers. The most efficient ones are those that are based on the principle of double refraction, to be treated in Chapter 6. Another type makes use of the phenomenon of anisotropic optical absorption, or *dichroism*, which means that one component of polarization is more strongly absorbed than the other. The natural crystal tourmaline is dichroic and can be used to make a polarizer, although it is not very efficient. A familiar commercial product is Polaroid, developed by Edwin Land. It consists of a thin layer of parallel needlelike crystals that are highly dichroic. The layer is embedded in a plastic sheet which can be cut and bent.

The *transmission axis* of such a polarizer defines the direction of the electric field vector for a light wave that is transmitted with little or no loss. A light wave whose electric vector is at right angles to the transmission axis is absorbed or attenuated. An ideal polarizer is one that is completely transparent to light linearly polarized in the direction of the transmission axis, and completely opaque to light polarized in the orthogonal direction to the transmission axis.

Consider the case of unpolarized light incident on an ideal linear polarizer. Now the instantaneous electric field E can always be resolved into two mutually perpendicular components, E_1 and E_2, (Figure 2.3), where E_1 is along the transmission axis of the polarizer.

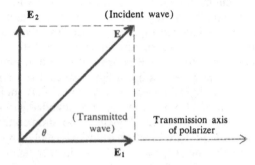

Figure 2.3. Relationship between the incident and the transmitted fields for a linear polarizer.

If E makes an angle θ with the transmission axis, then the magnitude of the transmitted field is

$$E_1 = E \cos \theta$$

The transmitted intensity I_1, being proportional to the square of the field, is therefore given by

$$I_1 = I \cos^2 \theta$$

where I is the intensity of the incident beam. For unpolarized light all values of θ occur with equal probability. Therefore, the transmission factor of an ideal linear polarizer for unpolarized light is just the average value of $\cos^2 \theta$ namely, $\frac{1}{2}$.

Partial Polarization   Light that is partially polarized can be considered to be a mixture of polarized and unpolarized light. The *degree of polarization* in this case is defined as the fraction of the total intensity that is polarized:

$$P = \text{degree of polarization} = \frac{I_{pol}}{I_{pol} + I_{unpol}} \qquad (2.26)$$

It is left as an exercise to show that for partial linear polarization

$$P = \frac{I_{max} - I_{min}}{I_{max} + I_{min}} \qquad (2.27)$$

where I_{max} and I_{min} refer to the intensity of the light transmitted through a linear polarizer when it is turned through the complete range of 360 degrees.

Scattering and Polarization   When light propagates through a medium other than a vacuum, the electric field of the light wave induces oscillating electric dipoles in the constituent atoms and molecules of the medium. It is these induced dipoles that are mainly responsible for the optical properties of a given substance, that is, refraction, absorption, and so on. This subject will be treated later in Chapter 6. In addition to affecting the propagation of light waves, the induced dipoles can also scatter the light in various directions. This molecular scattering (as distinguished from scattering by suspended particles such as dust) was investigated by Lord Rayleigh who showed that, theoretically, the fraction of light scattered by gas molecules should be proportional to the fourth power of the light frequency or, equivalently, to the inverse fourth power of the wavelength. This accounts for the blue color of the sky since the shorter wavelengths (blue region of the spectrum) are scattered more than the longer wavelengths (red region).[3]

In addition to the wavelength dependence of light scattering, there is also a polarization effect. This comes about from the directional radiation pattern of an oscillating electric dipole. The maximum radiation is emitted at right angles to the dipole axis, and no radiation is emitted along the direction of the axis. Furthermore the radiation is linearly polarized along the direction of the dipole axis. Consider the case of light that is scattered through an angle of 90 degrees. The electric vector of the scattered wave will be at right angles to the direction of the incident wave, as shown in Figure 2.4, and so the scattered light is linearly polarized. The polarization of the light of the blue sky is easily observed with a piece of Polaroid. The maximum amount of polarization is found in a direction 90 degrees from the direction of the sun. Measurements show that the degree of polarization can be greater than 50 percent.

2.4   Circular and Elliptic Polarization

Let us return temporarily to the real representation for electromagnetic waves. Consider the special case of two linearly polarized

---

[3] Actually, the sky would appear violet rather than blue were it not for the fact that the color sensitivity of the eye drops off sharply at the violet end of the spectrum and also that the energy in the solar spectrum diminishes there.

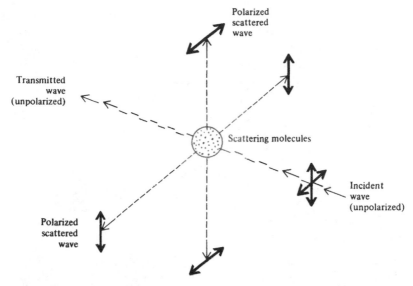

Figure 2.4. Illustrating polarization in the molecular scattering of light. The E vectors for the incident and scattered waves are indicated.

waves of the same amplitude E_0 polarized orthogonally to each other. Further, suppose the waves have a phase difference of $\pi/2$. We choose coordinate axes such that the electric vectors of the two waves are in the x and y directions, respectively. Accordingly, the component electric fields are

$$\hat{\mathbf{i}}E_0 \cos (kz - \omega t)$$

$$\hat{\mathbf{j}}E_0 \sin (kz - \omega t)$$

The total electric field **E** is the vector sum of the two component fields, namely,

$$\mathbf{E} = E_0 \left[\hat{\mathbf{i}} \cos (kz - \omega t) + \hat{\mathbf{j}} \sin (kz - \omega t) \right] \qquad (2.28)$$

Now the above expression is a perfectly good solution of the wave equation. It can be interpreted as a single wave in which the electric vector at a given point is constant in magnitude but rotates with angular frequency ω. This type of wave is said to be *circularly polarized*. A drawing showing the electric field and associated magnetic field of circularly polarized waves is shown in Figure 2.5.

The signs of the terms in Equation (2.28) are such that the expression represents *clockwise* rotation of the electric vector *at a given point in space* when viewed against the direction of propagation. Also, *at a given instant in time,* the field vectors describe right-

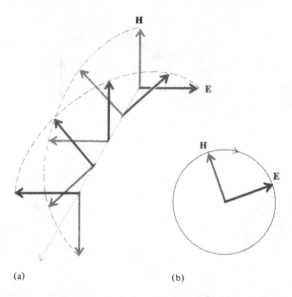

Figure 2.5.   Electric and magnetic vectors for right circularly polarized light. (a) Vectors at a given instant in time; (b) rotation of the vectors at a given position in space.

handed spirals as illustrated in Figure 2.5. Such a wave is said to be *right circularly polarized*.

If the sign of the second term is changed, then the sense of rotation is changed. In this case the rotation is counterclockwise at a given point in space when viewed against the direction of propagation, and, at a given instant in time the fields describe left-handed spirals. The wave is then called *left circularly polarized*.

It should perhaps be pointed out here that if one "rides along" with the wave, then the field vectors do not change in either direction or magnitude, because the quantity $kz - \omega t$ remains constant. This is true for any type of polarization.

Let us now return to the complex notation. The electric field for a circularly polarized wave can be written in complex form as

$$\mathbf{E} = \hat{\mathbf{i}}E_0 \exp i(kz - \omega t) + \hat{\mathbf{j}}E_0 \exp i(kz - \omega t \pm \pi/2) \qquad (2.29)$$

or, by employing the identity $e^{i\pi/2} = i$, we can write

$$\mathbf{E} = E_0(\hat{\mathbf{i}} \pm i\hat{\mathbf{j}}) \exp i(kz - \omega t) \qquad (2.30)$$

It is easy to verify that the real part of the above expression is precisely that of Equation (2.28) where, however, the minus sign must be used to represent right circular polarization and the plus sign for left circular polarization.

The reader is reminded here that if one uses the wave function $\exp i(\omega t - kz)$ rather than $\exp i(kz - \omega t)$, then the opposite sign convention applies.

Elliptic Polarization   If the component (real) fields are not of the same amplitude, say $\hat{\mathbf{i}}E_0 \cos (kz - \omega t)$ and $\hat{\mathbf{j}}E_0' \sin (kz - \omega t)$ where $E_0 \neq E_0'$, the resultant electric vector, at a given point in space, rotates and also changes in magnitude in such a manner that the end of the vector describes an ellipse as illustrated in Figure 2.6. In this case the wave is said to be *elliptically polarized*.

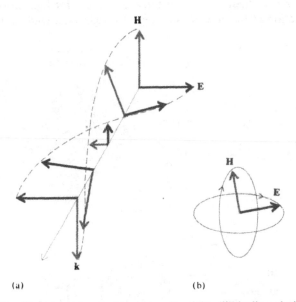

(a)   (b)

Figure 2.6.   Electric and magnetic vectors for right elliptically polarized light. (a) Vectors at a given instant in time; (b) at a given position in space.

It is sometimes convenient to employ a *complex vector amplitude* $\mathbf{E}_0$ defined as follows:

$$\mathbf{E}_0 = \hat{\mathbf{i}}E_0 + i\hat{\mathbf{j}}E_0' \tag{2.31}$$

The corresponding wave function is

$$\mathbf{E} = \mathbf{E}_0 \exp i(kz - \omega t) \tag{2.32}$$

This expression can represent any type of polarization. Thus if $\mathbf{E}_0$ is real, we have linear polarization, whereas if it is complex, we have elliptic polarization. In the special case of circular polarization the real and imaginary parts of $\mathbf{E}_0$ are equal.

Quarter-Wave Plate   Circularly polarized light can be produced by introducing a phase shift of $\pi/2$ between two orthogonal components of linearly polarized light. One device for doing this is known as a *quarter-wave plate*. These plates are made of *doubly refracting* transparent crystals, such as calcite or mica.[4] Doubly refracting crystals have the property that the index of refraction differs for different directions of polarization. It is possible to cut a doubly refracting crystal into slabs in such a way that an axis of maximum index n_1 (the slow axis) and an axis of minimum index n_2 (the fast axis) both lie at right angles to one another in the plane of the slab. If the slab thickness is d, then the optical thickness is $n_1 d$ for light polarized in the direction of the slow axis and $n_2 d$ for light polarized in the direc-

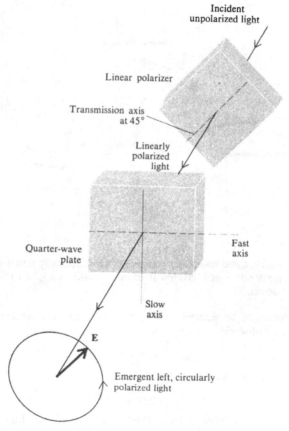

Figure 2.7.   Arrangement for producing circularly polarized light.

[4] The optics of crystals will be treated in detail in Chapter 6.

tion of the fast axis. For a quarter-wave plate, d is chosen to make the difference $n_1d - n_2d$ equal to one-quarter wavelength, so that d is given by the equation

$$d = \frac{\lambda_0}{4(n_1 - n_2)} \tag{2.33}$$

in which λ_0 is the vacuum wavelength.

The physical arrangement for producing circularly polarized light is shown in Figure 2.7. Incident unpolarized light is made linearly polarized by means of a linear polarizer such as a sheet of Polaroid. The quarter-wave plate is placed in the beam of linearly polarized light. The orientation of the quarter-wave plate is defined by the angle θ between the transmission axis of the Polaroid and the fast axis of the quarter-wave plate. By choosing θ to be 45 degrees, the light entering the quarter-wave plate can be resolved into two orthogonal linearly polarized components of equal amplitude and equal phase. On emerging from the quarter-wave plate, these two components are out of phase by $\pi/2$. Hence the emerging light is circularly polarized.

The sense of rotation of the circularly polarized light depends on the value of θ and can be reversed by rotating the quarter-wave plate through an angle of 90 degrees so that θ is 135 degrees. If θ is any value other than ± 45 degrees or ± 135 degrees, the polarization of the emerging light will be elliptic rather than circular.

2.5 Matrix Representation of Polarization. The Jones Calculus

The complex vector amplitude given in the preceding section, Equation (2.31), is not the most general expression, because it was assumed that the x component was real and the y component was imaginary. A more general way of expressing the complex amplitude of a plane harmonic wave is

$$\mathbf{E}_0 = \hat{\mathbf{i}}E_{0x} + \hat{\mathbf{j}}E_{0y} \tag{2.34}$$

where E_{0x} and E_{0y} can *both* be complex. Accordingly, they can be written in exponential form as

$$E_{0x} = |E_{0x}|\, e^{i\phi_x} \tag{2.35}$$

$$E_{0y} = |E_{0y}|\, e^{i\phi_y} \tag{2.36}$$

A convenient notation for the above pair of complex amplitudes is the following matrix known as the *Jones vector:*

$$\begin{bmatrix} E_{0x} \\ E_{0y} \end{bmatrix} = \begin{bmatrix} |E_{0x}|\, e^{i\phi_x} \\ |E_{0y}|\, e^{i\phi_y} \end{bmatrix} \tag{2.37}$$

The *normalized* form of the Jones vector is obtained by dividing by the square root of the sum of the squares of the two moduli, namely $(|E_{0x}|^2 + |E_{0y}|^2)^{1/2}$. A useful but not necessarily normalized form is obtained by factoring out any common factor that results in the simplest expression. For example, $\begin{bmatrix} A \\ 0 \end{bmatrix} = A \begin{bmatrix} 1 \\ 0 \end{bmatrix}$ represents a wave linearly polarized in the x direction, and $\begin{bmatrix} 0 \\ A \end{bmatrix} = A \begin{bmatrix} 0 \\ 1 \end{bmatrix}$ a wave linearly polarized in the y direction. The vector $\begin{bmatrix} A \\ A \end{bmatrix} = A \begin{bmatrix} 1 \\ 1 \end{bmatrix}$ represents a wave that is linearly polarized at 45 degrees relative to the x axis. Circular polarization is represented by $\begin{bmatrix} 1 \\ i \end{bmatrix}$ for left circular polarization, and $\begin{bmatrix} 1 \\ -i \end{bmatrix}$ for right circular polarization.

One of the applications of the Jones notation is calculating the result of adding two or more waves of given polarizations. The result is obtained simply by adding the Jones vectors. As an example, suppose we want to know the result of adding two waves of equal amplitude, one being right circularly polarized, the other left circularly polarized. The calculation by means of the Jones vectors proceeds as follows:

$$\begin{bmatrix} 1 \\ -i \end{bmatrix} + \begin{bmatrix} 1 \\ i \end{bmatrix} = \begin{bmatrix} 1 + 1 \\ -i + i \end{bmatrix} = \begin{bmatrix} 2 \\ 0 \end{bmatrix} = 2 \begin{bmatrix} 1 \\ 0 \end{bmatrix}$$

The last expression shows that the resultant wave is linearly polarized in the x direction and its amplitude is twice that of either of the circular components.

Another use of the matrix notation is that of computing the effect of inserting a linear optical element, or a train of such elements, into a beam of light of given polarization. The optical elements are represented by 2×2 matrices called *Jones matrices*. The types of optical devices that can be so represented include linear polarizers, circular polarizers, phase retarders (quarter-wave plates, and so forth), isotropic phase changers, and isotropic absorbers. We give, without proof, the matrices for several optical elements in Table 2.1 [39].

The matrices are used as follows. Let the vector of the incident light be $\begin{bmatrix} A \\ B \end{bmatrix}$ and the vector of the emerging light be $\begin{bmatrix} A' \\ B' \end{bmatrix}$. Then

$$\begin{bmatrix} a & b \\ c & d \end{bmatrix} \begin{bmatrix} A \\ B \end{bmatrix} = \begin{bmatrix} A' \\ B' \end{bmatrix} \tag{2.38}$$

where $\begin{bmatrix} a & b \\ c & d \end{bmatrix}$ is the Jones matrix of the optical element. If light is sent

Table 2.1. JONES MATRICES FOR SOME LINEAR OPTICAL ELEMENTS

Optical Element ———————————————— *Jones Matrix*

Linear polarizer	Transmission axis horizontal	$\begin{bmatrix} 1 & 0 \\ 0 & 0 \end{bmatrix}$
	Transmission axis vertical	$\begin{bmatrix} 0 & 0 \\ 0 & 1 \end{bmatrix}$
	Transmission axis at $\pm 45°$	$\dfrac{1}{2}\begin{bmatrix} 1 & \pm1 \\ \pm1 & 1 \end{bmatrix}$
Quarter-wave plate	Fast axis vertical	$\begin{bmatrix} 1 & 0 \\ 0 & -i \end{bmatrix}$
	Fast axis horizontal	$\begin{bmatrix} 1 & 0 \\ 0 & i \end{bmatrix}$
	Fast axis at $\pm 45°$	$\dfrac{1}{\sqrt{2}}\begin{bmatrix} 1 & \pm i \\ \pm i & 1 \end{bmatrix}$
Half-wave plate	Fast axis either vertical or horizontal	$\begin{bmatrix} 1 & 0 \\ 0 & -1 \end{bmatrix}$
Isotropic phase retarder		$\begin{bmatrix} e^{i\phi} & 0 \\ 0 & e^{i\phi} \end{bmatrix}$
Relative phase changer		$\begin{bmatrix} e^{i\phi_x} & 0 \\ 0 & e^{i\phi_y} \end{bmatrix}$
Circular polarizer	Right	$\dfrac{1}{2}\begin{bmatrix} 1 & i \\ -i & 1 \end{bmatrix}$
	Left	$\dfrac{1}{2}\begin{bmatrix} 1 & -i \\ i & 1 \end{bmatrix}$

Note: Normalization factors are included in the table. These factors are necessary for energy considerations only and can be omitted in calculations concerned primarily with type of polarization. Also, the signs of all matrix elements containing the factor i should be changed if one uses the wave function $\exp i(\omega t - kz)$ rather than $\exp i(kz - \omega t)$.

through a train of optical elements, then the result is given by matrix multiplication:

$$\begin{bmatrix} a_n & b_n \\ c_n & d_n \end{bmatrix} \cdots \begin{bmatrix} a_2 & b_2 \\ c_2 & d_2 \end{bmatrix}\begin{bmatrix} a_1 & b_1 \\ c_1 & d_1 \end{bmatrix}\begin{bmatrix} A \\ B \end{bmatrix} = \begin{bmatrix} A' \\ B' \end{bmatrix} \tag{2.39}$$

To illustrate, suppose a quarter-wave plate is inserted into a beam of linearly polarized light as shown in Figure 2.6. Here the incoming beam is polarized at 45 degrees with respect to the horizontal (x axis), so that its vector, aside from an amplitude factor, is $\begin{bmatrix} 1 \\ 1 \end{bmatrix}$. From the table, the Jones matrix for a quarter-wave plate with the fast-axis horizontal is $\begin{bmatrix} 1 & 0 \\ 0 & i \end{bmatrix}$. The vector of the emerging beam is then given by

$$\begin{bmatrix} 1 & 0 \\ 0 & i \end{bmatrix}\begin{bmatrix} 1 \\ 1 \end{bmatrix} = \begin{bmatrix} 1 \\ i \end{bmatrix}$$

The emergent light is therefore left circularly polarized.

It should be noted that the Jones calculus is of use only for computing results with light that is initially polarized in some way. There is no Jones vector representation for unpolarized light.

Orthogonal Polarization   Two waves whose states of polarization are represented by the complex vector amplitudes $\mathbf{E}_1$ and $\mathbf{E}_2$ are said to be *orthogonally polarized* if

$$\mathbf{E}_1 \cdot \mathbf{E}_2^* = 0$$

where the asterisk denotes the complex conjugate.

For linearly polarized light, orthogonality merely means that the fields are polarized at right angles to one another. In the case of circular polarization it is readily seen that right circular and left circular polarizations are mutually orthogonal states. But, there is a corresponding orthogonal polarization for any type of polarization.

In terms of Jones vectors it is easy to verify that $\begin{bmatrix} A_1 \\ B_1 \end{bmatrix}$ and $\begin{bmatrix} A_2 \\ B_2 \end{bmatrix}$ are orthogonal if

$$A_1 A_2^* + B_1 B_2^* = 0 \qquad \text{(2.40)}$$

Thus, for example, $\begin{bmatrix} 2 \\ i \end{bmatrix}$ and $\begin{bmatrix} 1 \\ -2i \end{bmatrix}$ represent a particular pair of orthogonal states of elliptic polarization. These are shown in Figure 2.8.

It is instructive to note that light of arbitrary polarization can always be resolved into two orthogonal components. Thus resolution into linear components is written

$$\begin{bmatrix} A \\ B \end{bmatrix} = A \begin{bmatrix} 1 \\ 0 \end{bmatrix} + B \begin{bmatrix} 0 \\ 1 \end{bmatrix}$$

and into circular components is written

$$\begin{bmatrix} A \\ B \end{bmatrix} = \frac{1}{2}(A + iB)\begin{bmatrix} 1 \\ -i \end{bmatrix} + \frac{1}{2}(A - iB)\begin{bmatrix} 1 \\ i \end{bmatrix}$$

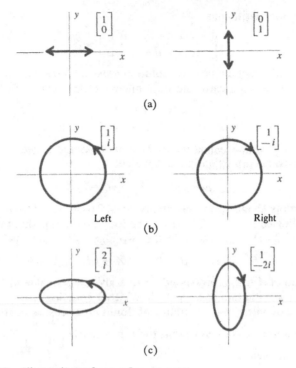

(a)

Left   (b)   Right

(c)

Figure 2.8. Illustrations of some Jones vectors.

Eigenvectors of Jones Matrices. An *eigenvector* of any matrix is defined as a particular vector which, when multiplied by the matrix, gives the same vector within a constant factor. In the Jones calculus this can be written

$$\begin{bmatrix} a & b \\ c & d \end{bmatrix}\begin{bmatrix} A \\ B \end{bmatrix} = \lambda \begin{bmatrix} A \\ B \end{bmatrix}$$

The constant λ, which may be real or complex, is called the *eigenvalue*.

Physically, an eigenvector of a given Jones matrix represents a particular polarization of a wave which, upon passing through the optical element in question, emerges with the same polarization as when it entered. However, depending on the value of λ, the amplitude and the phase may change. If we write $\lambda = |\lambda|e^{i\psi}$, then $|\lambda|$ is the amplitude change, and ψ is the phase change.

The problem of finding the eigenvalues and the corresponding eigenvectors of a 2×2 matrix is quite simple. The matrix equation above can be written as

$$\begin{bmatrix} a - \lambda & b \\ c & d - \lambda \end{bmatrix}\begin{bmatrix} A \\ B \end{bmatrix} = 0 \tag{2.41}$$

Now in order that a nontrivial solution exists, namely one in which A and B are not both zero, the determinant of the matrix must vanish

$$\begin{vmatrix} a - \lambda & b \\ c & d - \lambda \end{vmatrix} = 0 \tag{2.42}$$

This is a quadratic equation in λ, known as the *secular equation*. Upon expanding the determinant we get

$$(a - \lambda)(d - \lambda) - bc = 0$$

whose roots λ_1 and λ_2 are the eigenvalues. To each root there is a corresponding eigenvector. These can be found by noting that the matrix equation (2.41) is equivalent to the two algebraic equations

$$(a - \lambda)A + bB = 0 \qquad cA + (d - \lambda)B = 0 \tag{2.43}$$

The ratio of A to B, corresponding to a given eigenvalue of λ, can be found by substitution of λ_1 or λ_2 into either equation.

For example, from the table of Jones matrices, a quarter-wave plate with fast-axis horizontal has the Jones matrix $\begin{bmatrix} 1 & 0 \\ 0 & i \end{bmatrix}$. The secular equation is then

$$(1 - \lambda)(i - \lambda) = 0$$

which gives $\lambda = 1$ and $\lambda = i$ for the two eigenvalues. Equations (2.43) then read $(1 - \lambda)A = 0$ and $(i - \lambda)B = 0$. Thus, for $\lambda = 1$ we must have $A \neq 0$ and $B = 0$. Similarly, for $\lambda = i$ it is necessary that $A = 0$ and $B \neq 0$. Hence the normalized eigenvectors are $\begin{bmatrix} 1 \\ 0 \end{bmatrix}$ for $\lambda = 1$, and $\begin{bmatrix} 0 \\ 1 \end{bmatrix}$ for $\lambda = i$. Physically, the result means that light that is linearly polarized in the direction of either the fast axis or the slow axis is transmitted without change of polarization. There is no change in amplitude since $|\lambda| = 1$ for both cases, but a relative phase change of $\pi/2$ occurs since $\lambda_2/\lambda_1 = i = e^{i\pi/2}$.

2.6   Reflection and Refraction at a Plane Boundary

We now investigate the very basic phenomena of reflection and refraction of light from the standpoint of electromagnetic theory. It is assumed that the reader is already familiar with the elementary rules of reflection and refraction and how they are deduced from Huygens' principle. As we shall see, these rules can also be deduced from the application of boundary conditions for electromagnetic waves.

Consider a plane harmonic wave incident upon a plane boundary separating two different optical media (Figure 2.9). There will be a

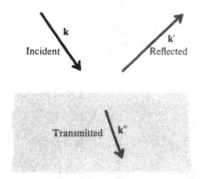

Figure 2.9. Wave vectors for light incident on a boundary separating two different optical media.

reflected wave and a transmitted wave. The space–time dependence of these three waves, aside from constant amplitude factors, is given by the following complex expressions:

$$\exp i(\mathbf{k} \cdot \mathbf{r} - \omega t) \qquad \text{incident wave}$$
$$\exp i(\mathbf{k}' \cdot \mathbf{r} - \omega t) \qquad \text{reflected wave}$$
$$\exp i(\mathbf{k}'' \cdot \mathbf{r} - \omega t) \qquad \text{transmitted (refracted) wave}$$

Now, in order that any constant relation can exist for all points of the boundary and for all values of t, it is necessary that the arguments of the three exponential functions be equal at the boundary. Thus, since the time factors are already equal, we must have

$$\mathbf{k} \cdot \mathbf{r} = \mathbf{k}' \cdot \mathbf{r} = \mathbf{k}'' \cdot \mathbf{r} \qquad \text{(at boundary)} \qquad (2.44)$$

These equations imply that all three wave vectors $\mathbf{k}$, $\mathbf{k}'$, and $\mathbf{k}''$ are coplanar, and that their projections onto the boundary plane are all equal. This can be argued by choosing a coordinate system $Oxyz$ such that one of the coordinate planes, say the xz plane, is the boundary, and also such that the vector $\mathbf{k}$ lies in the xy plane, called the *plane of incidence* as shown in Figure 2.10. The angles between the boundary normal (y axis) and the wave vectors are labeled θ, θ', and ϕ, as shown. Equation (2.44) then becomes

$$k \sin \theta = k' \sin \theta' = k'' \sin \phi \qquad (2.45)$$

Now in the space of the incident and reflected waves ($y > 0$), the two waves are traveling in the same medium, hence the wave vectors have

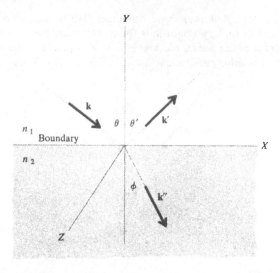

Figure 2.10. Coordinate system for analyzing reflection and refraction at a plane boundary.

the same magnitude; that is, $k = k'$. The first equation then reduces to the familiar law of reflection

$$\theta = \theta' \qquad (2.46)$$

Taking the ratio of the propagation constants of the transmitted wave and the incident wave, we have

$$\frac{k''}{k} = \frac{\omega/u''}{\omega/u} = \frac{c/u''}{c/u} = \frac{n_2}{n_1} = n \qquad (2.47)$$

where n_1 and n_2 are the indices of refraction of the two media, and n is *the relative index of refraction.* The second part of Equation (2.45) therefore is equivalent to Snell's law of refraction,

$$\frac{\sin \theta}{\sin \phi} = n \qquad (2.48)$$

2.7   Amplitudes of Reflected and Refracted Waves.

Let $\mathbf{E}$ denote the amplitude of the electric vector of a plane harmonic wave that is incident on a plane boundary separating two media, and let $\mathbf{E}'$ and $\mathbf{E}''$ denote the amplitudes of the reflected and transmitted waves, respectively. Then it follows from the Maxwell curl equations as applied to harmonic waves, Equation (2.11), that the corresponding amplitudes of the magnetic vectors are given by

$$\mathbf{H} = \frac{1}{\mu\omega} \mathbf{k} \times \mathbf{E} \qquad \text{(incident)} \qquad (2.49)$$

$$\mathbf{H}' = \frac{1}{\mu\omega} \mathbf{k}' \times \mathbf{E}' \qquad \text{(reflected)} \qquad (2.50)$$

$$\mathbf{H}'' = \frac{1}{\mu\omega} \mathbf{k}'' \times \mathbf{E}'' \qquad \text{(transmitted)} \qquad (2.51)$$

It should be noted that the above equations apply either to the instantaneous values of the fields or to the amplitudes, since the exponential factors exp $i(\mathbf{k} \cdot \mathbf{r} - \omega t)$, and so forth, are common to both the electric fields and associated magnetic fields.

It is convenient at this point to consider two different cases. The first case is that in which the electric vector of the incident wave is parallel to the boundary plane, that is, perpendicular to the plane of incidence. This case is called *transverse electric* or *TE* polarization. The second case is that in which the magnetic vector of the incident wave is parallel to the boundary plane. This is called *transverse magnetic* or *TM* polarization. The general case is handled by using appropriate linear combinations. The directions of the electric and magnetic vectors for the two cases are shown in Figure 2.11. As shown in the figure the boundary is taken to be the xz plane so that the y axis is normal to the boundary. The xy plane is the plane of incidence.

We now apply the well-known boundary conditions [16] which require that the tangential components of the electric and magnetic fields be continuous as the boundary is crossed. This means that for *TE* polarization $E + E' = E''$, and for *TM* polarization $H - H' = H''$. The results are as follows:

$$(TE \text{ polarization})$$
$$E + E' = E''$$
$$-H \cos \theta + H' \cos \theta = -H'' \cos \phi \qquad (2.52)$$
$$-kE \cos \theta + k'E' \cos \theta = -k''E'' \cos \phi$$

$$(TM \text{ polarization})$$
$$H - H' = H''$$
$$kE - k'E' = k''E'' \qquad (2.53)$$
$$E \cos \theta + E' \cos \theta = E'' \cos \phi$$

Here we have used the fact that each magnetic field amplitude H, H', and H'' is proportional to kE, $k'E'$, and $k''E''$, respectively, as implied by Equations (2.49) to (2.51).

The *coefficients of reflection* r_s and r_p, and the *coefficients of transmission* t_s and t_p are defined as amplitude ratios, namely

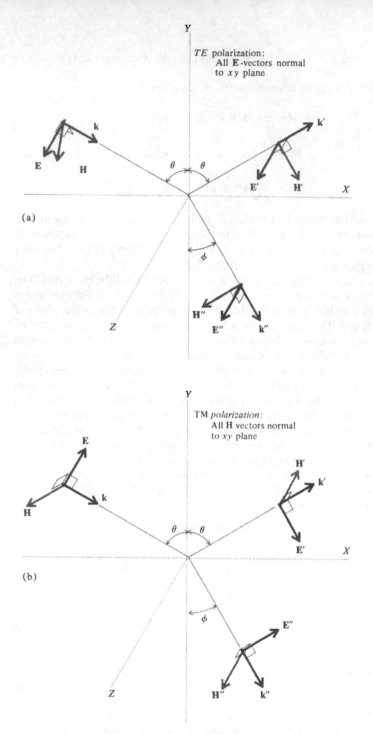

Figure 2.11. Wave vectors and associated fields for (a) *TE* and (b) *TM* polarization.

$$r_s = \left[\frac{E'}{E}\right]_{TE} \qquad r_p = \left[\frac{E'}{E}\right]_{TM}$$

$$t_s = \left[\frac{E''}{E}\right]_{TE} \qquad t_p = \left[\frac{E''}{E}\right]_{TM}$$

We next eliminate E'' from the two sets of equations (2.52) and (2.53) and use the relation $n = c/u = ck/\omega$ to obtain the following relations for the ratios of the reflected amplitudes to the incident amplitudes:

$$r_s = \frac{\cos\theta - n\cos\phi}{\cos\theta + n\cos\phi} \tag{2.54}$$

$$r_p = \frac{-n\cos\theta + \cos\phi}{n\cos\theta + \cos\phi} \tag{2.55}$$

Here

$$n = \frac{n_2}{n_1}$$

is the relative index of refraction of the two media. Ratios for the transmitted amplitudes can be obtained by eliminating E' in the two cases.

In the case of normal incidence both θ and ϕ are zero, and the expressions for r_s and r_p both reduce to the same thing, namely, $(1 - n)/(1 + n)$. The sign of this quantity is negative or positive depending on whether the relative index n is greater or less than unity, respectively. A negative value of E'/E means that the *phase* of the reflected wave is changed by 180 degrees relative to that of the incident wave. Thus such a change of phase occurs when light is partially reflected upon entering a dense medium, for example, from air to glass.

It should be noted here that some authors choose to define the positive sense for the **E'** and **H'** vectors of the reflected wave opposite to that shown in Figure 2.11(b) for the *TM* case. This leads to the awkward situation in which the *TE* and *TM* vectors must be defined differently for normal incidence, whereas there is actually no physical difference in this case.

Fresnel's Equations   By the use of Snell's law, $n = \sin\theta/\sin\phi$, the equations for the amplitudes of the reflected and refracted waves can be expressed in the form

$$r_s = -\frac{\sin(\theta - \phi)}{\sin(\theta + \phi)}$$

$$t_s = \frac{2\cos\theta\sin\phi}{\sin(\theta + \phi)}$$

(2.56)

$$r_p = -\frac{\tan(\theta - \phi)}{\tan(\theta + \phi)}$$

$$t_p = \frac{2\cos\theta\sin\phi}{\sin(\theta + \phi)\cos(\theta - \phi)}$$

(2.57)

The above equations are known as *Fresnel's equations*. Their derivation is left as a problem.

A third way of expressing the amplitude ratios for reflected light is to eliminate the variable ϕ in Equations (2.54) and (2.55) by use of Snell's law. The result is

$$r_s = \frac{\cos\theta - \sqrt{n^2 - \sin^2\theta}}{\cos\theta + \sqrt{n^2 - \sin^2\theta}}$$

(2.58)

$$r_p = \frac{-n^2\cos\theta + \sqrt{n^2 - \sin^2\theta}}{n^2\cos\theta + \sqrt{n^2 - \sin^2\theta}}$$

(2.59)

The *reflectance* is defined as the fraction of the incident light energy that is reflected, and is denoted by the symbols R_s and R_p for the *TE* and the *TM* case, respectively. Since the energy is proportional to the absolute square of the field amplitude, we have

$$R_s = |r_s|^2 = \left|\frac{E'}{E}\right|^2_{TE}$$

$$R_p = |r_p|^2 = \left|\frac{E'}{E}\right|^2_{TM}$$

(2.60)

Figure 2.12 shows the variation of E'/E and $|E'/E|^2$ with the angle of incidence for the two cases as calculated from the preceding theory.

For normal incidence ($\theta = 0$) we find that R_s and R_p reduce to the same value, namely

$$R_s = R_p = \left[\frac{n-1}{n+1}\right]^2$$

(2.61)

Thus, for glass of index 1.5 the reflectance at normal incidence is equal to $(0.5/2.5)^2 = 0.04$ for a single glass-air interface. In the case of an optical instrument, such as a camera, which may contain many lenses, this 4 percent reflective loss at each interface can build up to a substantial overall loss in the system. In order to reduce such losses, the surfaces of lenses and other optical elements are usually coated with antireflecting films, the theory of which is discussed in Chapter 4.

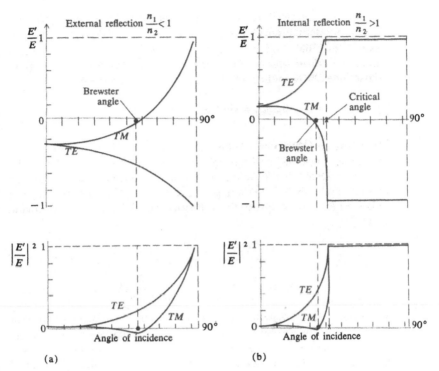

Figure 2.12. Graphs of E'/E and $|E'/E|^2$ versus angle of incidence for (a) external reflection and (b) internal reflection. (Approximate curves for glass of index 1.5.)

For grazing incidence ($\theta \sim 90$ degrees) the reflectance is also the same for both types of polarization, namely, unity, and it is independent of n. A smooth surface of any material is highly reflecting at grazing incidence.

External and Internal Reflection   In order to discuss the reflection of light for intermediate values of θ, we must distinguish between two different possibilities. These are first, the case in which the relative index of refraction $n_2/n_1 = n$ is greater than unity. This is called *external reflection*. In the second case n is less than unity. This is known as *internal reflection*. In external reflection the incident wave approaches the boundary from the side with the smaller index of refraction, whereas in internal reflection the incident wave is in the medium having the larger index of refraction.

In the case of external reflection, $n > 1$, the amplitude ratios as given by Equations (2.54) to (2.59) are real for all values of θ. The calculation of the reflectance R is then perfectly straightforward. For the case of internal reflection however, since $n < 1$, there will be values of θ such that $\sin \theta > n$, that is, $\theta > \sin^{-1} n$. The angle $\sin^{-1} n$ is called the *critical angle*. Thus the critical angle for ordinary glass whose index of refraction relative to air is 1.5 is given by

$$\theta_{critical} = \sin^{-1} \frac{1}{1.5} \approx 41 \text{ degrees}$$

Total Internal Reflection   When the internal angle of incidence exceeds the critical angle, the ratio E'/E becomes complex. This can be seen by referring to Equations (2.58) and (2.59) and noting that the quantity under the radical is negative for $\theta > \sin^{-1}n$. For this range of values of θ we can express the reflection coefficients in the following way:

$$r_s = \frac{\cos \theta - i \sqrt{\sin^2 \theta - n^2}}{\cos \theta + i \sqrt{\sin^2 \theta - n^2}} \qquad (2.62)$$

$$r_p = \frac{-n^2 \cos \theta + i \sqrt{\sin^2 \theta - n^2}}{n^2 \cos \theta + i \sqrt{\sin^2 \theta - n^2}} \qquad (2.63)$$

By multiplying by the complex conjugates, one can easily verify that the square of the absolute values of each of the above ratios is equal to unity. This means that $R = 1$; that is, we have *total reflection* when the internal angle of incidence is equal to or greater than the critical angle.

Fiber Optics and Optical Waveguides   One of the many practical applications of total internal reflection is the transmission of light through small continuous fibers (light guides). Bundles of such fibers can actually transmit images, and the bundles can be made flexible enough to withstand a certain amount of bending. If we consider a single fiber to be a solid dielectric cylinder immersed in a surrounding medium of index of refraction lower than that of the fiber, then light traveling in the general direction of the fiber axis will be trapped if the angle of incidence of a light ray on the wall of the fiber is equal to or greater than the critical angle. The *acceptance angle*, defined as the maximum semiangle of the vertex of a cone of rays entering the fiber from one end (Figure 2.13), is easily shown to be given by

$$\alpha = \sin^{-1} \sqrt{n_1^2 - n_2^2}$$

where n_1 and n_2 are the indices of refraction of the fiber and surrounding material, respectively (see Problem 2.20). In the case of very small uniform fibers, of the order of a few microns in diameter, monochromatic light travels through the fiber with a very definite electromagnetic wave pattern or *mode*. In this case we have an *op-*

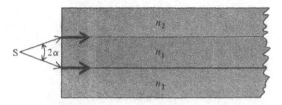

Figure 2.13. Geometry defining the acceptance angle of an optical waveguide. (The source S is considered to be in air, $n_0 = 1$.)

tical waveguide. Losses in such waveguides can be made very low. The device is potentially important in optical communications (via laser light), data processing, and many other applications.

2.8 The Brewster Angle

From Equation (2.59), which gives the amplitude ratio for reflection in the TM case, we see that the reflection is zero for that particular angle of incidence θ such that

$$\theta = \tan^{-1} n \tag{2.64}$$

This angle is called the *polarizing angle* or the *Brewster angle*. For example, for glass of index 1.5 we have for external reflection from air to glass

$$\theta_{\text{Brewster}} = \tan^{-1} 1.5 \approx 57 \text{ degrees} \tag{2.65}$$

and for internal reflection from glass to air

$$\theta_{\text{Brewster}} = \tan^{-1}\left(\frac{1}{1.5}\right) \approx 33 \text{ degrees} \tag{2.66}$$

Strictly speaking, the Brewster angle is a function of wavelength because of dispersion. The variation over the visible spectrum is very small, however.

If unpolarized light is incident on a surface at the Brewster angle, the reflected light is rendered linearly polarized with the electric vector transverse to the plane of incidence. The transmitted light is partially polarized. Although the reflected beam is completely polarized, only a small fraction of the light is reflected, namely, about 15 percent of the TE component for an air-glass interface, the index of the glass being 1.5. Hence the production of polarized light by reflection at the Brewster angle is not very efficient.

Brewster Window Suppose a beam of light that is linearly polarized in the TM mode is incident at Brewster's angle on a glass plate with parallel faces, as shown in Figure 2.14. Then no light is reflected from

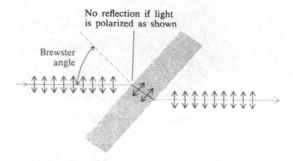

Figure 2.14. A Brewster window.

the first face. Neither is there any internal reflection from the second face. (The proof of this statement is left as a problem.) The result is that the light is entirely transmitted; in other words, we have a perfect window. Such devices, known as Brewster windows, are used extensively in laser applications.

Pile-of-Plates Polarizer   If unpolarized light is passed through a Brewster window, the transmitted light is partially polarized but with a fairly low degree of polarization. This can be increased by using a number of plates in tandem, called a "pile-of-plates" polarizer. The device is especially useful in the infrared region of the spectrum.

2.9   The Evanescent Wave in Total Reflection

In spite of the fact that the incident energy is totally reflected when the angle of incidence exceeds the critical angle, there is still an electromagnetic wave field in the region beyond the boundary. This field is known as the *evanescent wave*. Its existence can be understood by consideration of the wave function of the electric vector of the transmitted wave:

$$\mathbf{E}_{\text{trans}} = \mathbf{E}'' \, e^{i(\mathbf{k}'' \cdot \mathbf{r} - \omega t)}$$

On choosing coordinate axes as shown in Figure 2.10, we have

$$\mathbf{k}'' \cdot \mathbf{r} = k'' \, x \sin \phi - k'' y \cos \phi$$

$$= k'' x \sin \phi - i k'' y \sqrt{\frac{\sin^2 \theta}{n^2} - 1} \tag{2.65}$$

where in the last step we have used Snell's law in the following way:

$$\cos \phi = \sqrt{1 - \frac{\sin^2 \theta}{n^2}} = i \sqrt{\frac{\sin^2 \theta}{n^2} - 1} \tag{2.66}$$

The above equation shows that cos ϕ is imaginary in the case of total internal reflection. The wave function for the electric vector of the transmitted wave is then expressible as

$$\mathbf{E}_{\text{trans}} = \mathbf{E}'' e^{-\alpha|y|} e^{i(k_1 x - \omega t)} \tag{2.67}$$

where

$$\alpha = k'' \sqrt{\frac{\sin^2 \theta}{n^2} - 1}$$

and

$$k_1 = \frac{k'' \sin \theta}{n}$$

The factor $e^{-\alpha|y|}$ in Equation (2.67) shows that the evanescent wave amplitude drops off very rapidly as we proceed away from the boundary into the rarer medium. The complex exponential factor $e^{i(k_1 x - \omega t)}$ indicates that the evanescent wave can be described in terms of surfaces of constant phase moving parallel to the boundary with speed ω/k_1. It is easy to show that this is greater by the factor $1/\sin \theta$ than the phase velocity of ordinary plane waves in the denser medium.

That the wave actually penetrates into the rarer medium can be demonstrated experimentally in several ways. One way is shown in Figure 2.15. Two 45-90-45-degree prisms are placed with their long

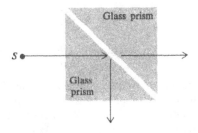

Figure 2.15. A method for illustrating the penetration of light into the rare medium.

faces close together but not in actual contact. Light from a source S is found to be partially transmitted, the amount depending on the separation of the prism faces. This arrangement can be used to make such things as variable output couplers for lasers. In another experiment, first demonstrated by Raman, reflection of light was found to occur from a sharp metallic edge placed near, but not in contact with, the edge of a totally reflecting prism as shown in Figure 2.16.

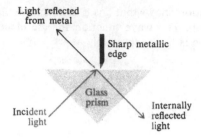

Figure 2.16.   Raman's demonstration that light penetrates into the rare medium.

In the absence of any perturbing influence in the rarer medium, the evanescent wave must return to the denser medium, because there is, in fact, total reflection of the light energy. In the case of a very narrow beam of light, the reflected beam is physically displaced from the geometrically predicted beam by a very small amount.[5]

2.10   Phase Changes in Total Internal Reflection

In the case of total internal reflection the complex values for the coefficients of reflection, given by Equations (2.62) and (2.63), imply that there is a change of phase which is a function of the angle of incidence. We now proceed to calculate this phase change.

Since the absolute values of r_s and r_p are both unity, we can write

$$r_s = e^{-i\delta_s} = \frac{ae^{-i\alpha}}{ae^{+i\alpha}} \tag{2.68}$$

$$r_p = -e^{-i\delta_p} = -\frac{be^{-i\beta}}{be^{+i\beta}} \tag{2.69}$$

in which δ_s and δ_p are the phase changes for the TE and TM cases, respectively. The complex numbers $ae^{-i\alpha}$ and $-be^{-i\beta}$ are equal to the numerators in Equations (2.62) and (2.63). Their complex conjugates appear in the denominators. Thus

$$ae^{i\alpha} = \cos\theta + i\sqrt{\sin^2\theta - n^2}$$

$$be^{i\beta} = n^2\cos\theta + i\sqrt{\sin^2\theta - n^2}$$

From Equation (2.68) above, we have $\delta_s = 2\alpha$. Accordingly, $\tan\alpha = \tan(\delta_s/2)$. Similarly, from Equation (2.69), $\tan\beta = \tan(\delta_p/2)$. We find, therefore, the following expressions for the phase changes that occur in internal reflection:

[5] This effect was studied by Goos and Haenchen in 1947 and is known as the *Goos-Haenchen shift*.

$$\tan \frac{\delta_s}{2} = \frac{\sqrt{\sin^2 \theta - n^2}}{\cos \theta} \tag{2.70}$$

$$\tan \frac{\delta_p}{2} = \frac{\sqrt{\sin^2 \theta - n^2}}{n^2 \cos \theta} \tag{2.71}$$

From these the relative phase difference can be found.

$$\Delta = \delta_p - \delta_s$$

By use of the appropriate trigonometric identities, it is found that the relative phase difference can be expressed as

$$\tan \frac{\Delta}{2} = \frac{\cos \theta \sqrt{\sin^2 \theta - n^2}}{\sin^2 \theta} \tag{2.72}$$

Graphs of δ_s and δ_p are shown in Figure 2.17. The curves illustrate how the phase change varies with the internal angle of incidence.

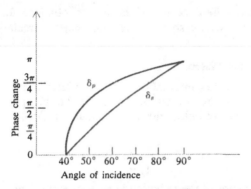

Figure 2.17. Phase changes occurring in total internal reflection. (Approximate curves for glass of index 1.5.)

Fresnel's Rhomb A method of changing linearly polarized light into circularly polarized light, devised by Fresnel, is shown in Figure 2.18. The essential element is a glass prism made in the form of a rhomb. Linearly polarized light whose polarization is oriented at an angle of 45 degrees with respect to the face edge of the rhomb enters normally on one face as shown. The light undergoes two internal reflections and emerges from the exit face. At each internal reflection a phase difference is produced between the component TM and TE polarizations. This phase difference Δ is equal to $\pi/4$ for an angle θ of 54 degrees (with glass of index 1.5) as calculated from Equation (2.72).

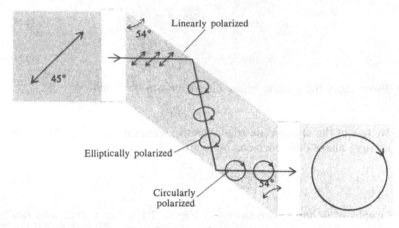

Figure 2.18. The Fresnel rhomb.

A total phase difference of $\pi/2$ is thereby obtained for the two internal reflections, and so the emerging light is circularly polarized.

2.11 Reflection Matrix

In the case of either internal or external reflection, if we identify the *TM* components as "horizontal" and the *TE* components as "vertical," then we can utilize the Jones calculus by defining a *reflection matrix*[6] as

$$\begin{bmatrix} -r_p & 0 \\ 0 & r_s \end{bmatrix}$$

The Jones vector of the reflected light is then given by

$$\begin{bmatrix} -r_p & 0 \\ 0 & r_s \end{bmatrix} \begin{bmatrix} A \\ B \end{bmatrix} = \begin{bmatrix} A' \\ B' \end{bmatrix} \tag{2.73}$$

where $\begin{bmatrix} A \\ B \end{bmatrix}$ is the vector of the incident light. The values of r_p and r_s are given as functions of the angle of incidence by Equations (2.58) and (2.59).

We can similarly define a *transmission matrix* as

$$\begin{bmatrix} t_p & 0 \\ 0 & t_s \end{bmatrix}$$

---

[6] Off-diagonal elements of these matrices are zero for isotropic dielectrics such as glass. Nonzero values of the off-diagonal elements can occur in the case of anisotropic substances, such as crystals.

and the Jones vector of the transmitted light is

$$\begin{bmatrix} t_p & 0 \\ 0 & t_s \end{bmatrix}\begin{bmatrix} A \\ B \end{bmatrix} = \begin{bmatrix} A'' \\ B'' \end{bmatrix} \tag{2.74}$$

However, we shall be mainly concerned with reflection in the present discussion.

Consider, for example, the case of normal incidence. The reflection matrix is then simply

$$\begin{bmatrix} -(1-n)/(1+n) & 0 \\ 0 & (1-n)/(1+n) \end{bmatrix} = \frac{1-n}{1+n}\begin{bmatrix} -1 & 0 \\ 0 & 1 \end{bmatrix}$$

in which n is the relative index of refraction. Suppose that the incident light is circularly polarized, say right. so that its Jones vector is $\begin{bmatrix} 1 \\ -i \end{bmatrix}$. The Jones vector of the reflected light is then

$$\frac{1-n}{1+n}\begin{bmatrix} -1 & 0 \\ 0 & 1 \end{bmatrix}\begin{bmatrix} 1 \\ -i \end{bmatrix} = \frac{1-n}{1+n}\begin{bmatrix} -1 \\ -i \end{bmatrix} = \frac{n-1}{n+1}\begin{bmatrix} 1 \\ i \end{bmatrix}$$

Thus the reflected light is left circularly polarized, and its amplitude is changed by the factor $(n-1)/(n+1)$. Similarly, it is found that if the incident light is left circularly polarized, then the reflected light is right circularly polarized. This reversal of the handedness of the circular polarization is independent of the value of n, and thus occurs in the case of both internal and external reflection provided the angle of incidence is small.

Next, consider the case of reflection near grazing incidence. Here r_p and r_s are *opposite* in sign and of magnitude unity. Hence the reflection matrix becomes

$$\pm\begin{bmatrix} 1 & 0 \\ 0 & 1 \end{bmatrix}$$

in which the $+$ sign applies for external reflection, and the $-$ sign for internal reflection. If the incident light is circularly polarized, then the sense of rotation is *not* changed upon reflection. This is true for both internal and external reflection.

Finally, let us consider the case of total internal reflection. Here, as we have shown, $r_p = -e^{-\delta_p}$ and $r_s = e^{-\delta_s}$. The reflection process is then simply written

$$\begin{bmatrix} A' \\ B' \end{bmatrix} = \begin{bmatrix} e^{-i\delta_p} & 0 \\ 0 & e^{-i\delta_s} \end{bmatrix}\begin{bmatrix} A \\ B \end{bmatrix} = \begin{bmatrix} Ae^{-i\delta_p} \\ Be^{-i\delta_s} \end{bmatrix} = e^{-i\delta_p}\begin{bmatrix} A \\ Be^{i\Delta} \end{bmatrix} \tag{2.75}$$

in which Δ is the relative phase difference given by Equation (2.72). The reflected light is, in general, elliptically polarized. See Problem 2.9.

PROBLEMS

2.1   Verify that the del operator applied to a plane harmonic wave function $f = e^{i(\mathbf{k}\cdot\mathbf{r}-\omega t)}$ yields the result $\nabla f = i\mathbf{k}f$.

2.2   What is the rms value of the electric field of the radiation from a 100-watt light bulb at a distance of 1 m?

2.3   The peak power of a ruby laser is 100 megawatts (MW). If the beam is focused to a spot 10 microns (μ) in diameter, find the irradiance and the amplitude of the electric field of the light wave at the focal point. The index of refraction is $n = 1$.

2.4   Show that the average Poynting flux is given by the expression $\frac{1}{2}Re\ (\mathbf{E}_0 \times \mathbf{H}_0^*)$, where $\mathbf{E}_0$ and $\mathbf{H}_0$ are the complex field amplitudes of a light wave.

2.5   The electric vector of a wave is given by the real expression

$$\mathbf{E} = E_0[\hat{\mathbf{i}} \cos (kz - \omega t) + \hat{\mathbf{j}}b \cos (kz - \omega t + \phi)]$$

Show that this is equivalent to the complex expression

$$\mathbf{E} = E_0(\hat{\mathbf{i}} + \hat{\mathbf{j}}be^{i\phi})e^{i(kz-\omega t)}$$

2.6   Sketch diagrams to show the type of polarization in Problem 2.5 for the following cases:
(a) $\phi = 0, b = 1$
(b) $\phi = 0, b = 2$
(c) $\phi = \pi/2, b = -1$
(d) $\phi = \pi/4, b = 1$

2.7   Write down the Jones vectors for the waves in Problem 2.6.

2.8   Describe the type of polarization of the waves whose Jones vectors are

$$\begin{bmatrix} 1 \\ \sqrt{3} \end{bmatrix}, \begin{bmatrix} i \\ -1 \end{bmatrix}, \begin{bmatrix} 1-i \\ 1+i \end{bmatrix}$$

Find the orthogonal Jones vectors to each of the above and describe their polarizations.

2.9   The general case is represented by the Jones vector

$$\begin{bmatrix} A \\ Be^{i\Delta} \end{bmatrix}$$

Show that this represents elliptically polarized light in which the major axis of the ellipse makes an angle

$$\frac{1}{2} \tan^{-1} \left(\frac{2AB \cos \Delta}{A^2 - B^2} \right)$$

with the x axis.

2.10   Show by means of the Jones calculus that circularly polarized light is produced by sending light through a linear polarizer and a quarter-wave plate *only* in the right order.

2.11 Verify that a circular polarizer whose Jones matrix is $\begin{bmatrix} 1 & i \\ -i & 1 \end{bmatrix}$ is completely transparent to one type of circularly polarized light and opaque to the opposite circular polarization. (Note: This is not the same type of polarizer as a linear polarizer followed by a quarter-wave plate.)

2.12 Linearly polarized light whose Jones vector is $\begin{bmatrix} 1 \\ 0 \end{bmatrix}$ (horizontally polarized) is sent through a train of two linear polarizers. The first is oriented with its transmission axis at 45 degrees and the second has its transmission axis vertical. Show that the emerging light is linearly polarized in the vertical direction; that is, the plane of polarization has been rotated by 90 degrees.

2.13 Find the eigenvalues and corresponding eigenvectors for a linear polarizer with its transmission axis at 45 degrees.

2.14 Find the critical angle for internal reflection in water ($n = 1.33$) and diamond ($n = 2.42$).

2.15 Find the Brewster angle for external reflection in water and diamond.

2.16 Find the reflectance for both TE and TM polarizations at an angle of incidence of 45 degrees for water and diamond.

2.17 The critical angle for total internal reflection in a certain substance is exactly 45 degrees. What is the Brewster angle for external reflection?

2.18 The accompanying figure (Figure 2.19) is a diagram of the Mooney rhomb for producing circularly polarized light. Show that the apex angle A should be about 60 degrees if the index of refraction of the rhomb is 1.65.

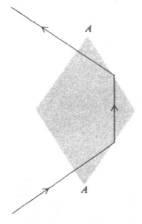

Figure 2.19. The Mooney rhomb.

2.19 A beam of light is totally reflected in a 45-90-45-degree glass prism ($n = 1.5$) as shown in Figure 2.15. The wavelength of the light is 500 nm. At what distance from the surface is the amplitude of the evanescent wave $1/e$ of its value at the surface? By what factor is the intensity of the evanescent wave reduced at a distance of 1 mm from the surface?

2.20 Show that the acceptance angle for a glass-fiber waveguide is given by

$$\alpha = \sin^{-1} \sqrt{n_1{}^2 - n_2{}^2}$$

where n_1 and n_2 are the indices of refraction of the fiber and the cladding material, respectively, and the external medium is air, $n_0 = 1$ (see Figure 2.13).

2.21 Fill in the steps leading to the equation

$$\tan \frac{\Delta}{2} = \frac{\cos \theta \sqrt{\sin^2 \theta - n^2}}{\sin^2 \theta}$$

for the phase difference in total internal reflection as discussed in Section 2.10.

2.22 A beam of light is right circularly polarized and is incident at an angle of 45 degrees on a glass surface ($n = 1.5$). Use the reflection matrix to find the polarization of the reflected light for both internal and external reflection.

2.23 Find the degree of polarization produced when a beam of unpolarized light passes through a Brewster window of index n. Find the numerical value for $n = 1.5$.

CHAPTER 3
Coherence and
Interference

3.1 The Principle of Linear Superposition

The theory of optical interference is based essentially on the principle of linear superposition of electromagnetic fields. According to this principle, the electric field E produced at a point in empty space jointly by several different sources is equal to the vector sum

$$E = E_{(1)} + E_{(2)} + E_{(3)} + \cdots \qquad (3.1)$$

where $E_{(1)}, E_{(2)}, E_{(3)}, \ldots$ are the fields produced at the point in question separately by the different sources. The same is true for magnetic fields. The principle is a consequence of the fact that Maxwell's equation for the vacuum are linear differential equations.

In the presence of matter, the principle of linear superposition is only approximately true. (This does not mean that fields cannot be resolved into components. It merely means that the resultant total field produced in matter by several external sources acting simultaneously may not be the same as the sum of the fields produced by each source acting alone.) Deviations from linearity are observed at the high intensities produced by lasers and come under the heading of nonlinear optical phenomena.[1]

Let us consider two plane harmonic linearly polarized waves of the same frequency ω. The electric fields are then

$$
\begin{aligned}
E_{(1)} &= E_1 \exp i(k_1 \cdot r - \omega t + \phi_1) \\
E_{(2)} &= E_2 \exp i(k_2 \cdot r - \omega t + \phi_2)
\end{aligned}
\qquad (3.2)
$$

Here the quantities ϕ_1 and ϕ_2 have been introduced to allow for any phase difference between the sources of the two waves. If the phase difference $\phi_1 - \phi_2$ is constant, the two sources are said to be *mutually coherent*. The resulting waves are also mutually coherent in this case.

For the present we shall confine our discussion to mutually co-

---

[1] Nonlinear optics will be treated later in Section 6.12.

herent monochromatic waves. The question of partial coherence and nonmonochromatic waves will be taken up later.

We know from Section 2.2 that the irradiance at a point is proportional to the square of the amplitude of the light field at the point in question. Thus the superposition of our two monochromatic plane waves, aside from a constant proportionality factor, results in an irradiance function

$$I = |\mathbf{E}|^2 = \mathbf{E} \cdot \mathbf{E}^* = (\mathbf{E}_{(1)} + \mathbf{E}_{(2)}) \cdot (\mathbf{E}^*_{(1)} + \mathbf{E}^*_{(2)})$$
$$= |\mathbf{E}_1|^2 + |\mathbf{E}_2|^2 + 2\mathbf{E}_1 \cdot \mathbf{E}_2 \cos \theta \qquad (3.3)$$
$$= I_1 + I_2 + 2\mathbf{E}_1 \cdot \mathbf{E}_2 \cos \theta$$

where

$$\theta = \mathbf{k}_1 \cdot \mathbf{r} - \mathbf{k}_2 \cdot \mathbf{r} + \phi_1 - \phi_2 \qquad (3.4)$$

The term $2\mathbf{E}_1 \cdot \mathbf{E}_2 \cos \theta$ is called the *interference term*. This term indicates I can be greater than or less than the sum $I_1 + I_2$, depending on the value of θ. Since θ depends on $\mathbf{r}$, periodic spatial variations in the intensity occur. These variations are the familiar interference fringes that are seen when two mutually coherent beams of light are combined.

If the sources of the two waves are mutually incoherent, then the quantity $\phi_1 - \phi_2$ varies with time in a random fashion. The result is that the mean value of $\cos \theta$ is zero, and there is no interference. This is the reason interference fringes are not observed with two separate (ordinary) light sources.

In the event that the two waves are polarized, then the interference term also depends on the polarization. In particular, if the polarizations are mutually orthogonal, then $\mathbf{E}_1 \cdot \mathbf{E}_2 = 0$. Again there are no interference fringes. This is true not only for linearly polarized waves but for circularly and elliptically polarized waves as well. The proof of the latter statement is left as a problem.

3.2 Young's Experiment

The classic experiment that demonstrates interference of light was first performed by Thomas Young in 1802. In the original experiment sunlight was used as the source, but any bright source such as a tungsten filament lamp or an arc would be satisfactory. Light is passed through a pinhole S so as to illuminate an aperture consisting of two pinholes or narrow slits S_1 and S_2 as shown in Figure 3.1. If a white screen is placed in the region beyond the slits, a pattern of bright and dark interference bands can be seen. The key to the experiment is the use of a single pinhole S to illuminate the aperture. This provides the necessary mutual coherence between the light coming from the two slits S_1 and S_2.

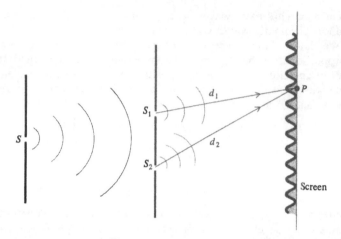

Figure 3.1. Young's experiment.

The usual elementary analysis of the Young experiment involves finding the difference in phase between the two waves arriving at a given point P over the distances d_1 and d_2 as shown. Assuming that we have spherical waves whose phase factors are of the form $e^{i(kr - \omega t)}$ as discussed in Section 1.4, the phase difference at P will be $k(d_2 - d_1)$. Bright fringes occur when this difference is 0, $\pm 2\pi$, $\pm 4\pi$, ... $\pm 2n\pi$, where n is a integer. These values give maxima for the interference term in Equation (3.3). Since $k = 2\pi/\lambda$, we see that the equation

$$k(d_2 - d_1) = \pm 2n\pi \tag{3.5}$$

is equivalent to

$$|d_2 - d_1| = n\lambda \tag{3.6}$$

that is, the path difference is equal to an integral number of wavelengths.

The result above can be related to the physical parameters of the experimental arrangement by labeling the geometry as shown in Figure 3.2. Here h is the slit separation and x is the distance from the slit aperture to the screen. The distance y is measured on the screen from the central axis as shown. Equation (3.6) is then equivalent to

$$\left[x^2 + \left(y + \frac{h}{2}\right)^2\right]^{1/2} - \left[x^2 + \left(y - \frac{h}{2}\right)^2\right]^{1/2} = n\lambda \tag{3.7}$$

By use of the binomial expansion we find that an approximate equivalent expression is given by

$$\frac{yh}{x} = n\lambda \tag{3.8}$$

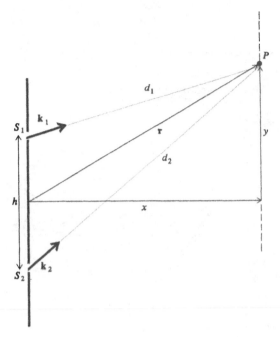

Figure 3.2. Geometry for analyzing interference in the case of the double slit.

The approximation is valid when y and h are both small compared to x. Bright fringes occur at the points

$$y = 0, \pm \frac{\lambda x}{h}, \pm \frac{2\lambda x}{h}, \cdots \qquad (3.9)$$

If the slits are covered by optical devices such as phase retarders, polarizers, and so forth, the fringe pattern will change. For example, if a relative phase difference of π is introduced by placing a thin plate of glass over one slit, then the whole interference pattern would shift by one half the fringe separation, so that a bright fringe would occur where a dark fringe was located before. The required thickness of the glass plate in this example is left as a problem for the student to calculate. Another important point is that if separate polarizers are placed over the two slits and oriented in such a manner that the component waves are orthogonally polarized, then $\mathbf{E}_1 \cdot \mathbf{E}_2 = 0$, hence no interference fringes occur.

Alternative Methods of Demonstrating Interference Other experimental arrangements for demonstrating interference between two waves are illustrated in Figure 3.3. All of these make use of reflection or refraction to obtain two mutually coherent waves that originate from a single source.

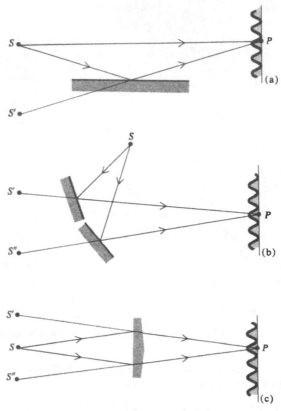

Figure 3.3. Arrangements for producing interference fringes with a single source. (a) Lloyd's single mirror; (b) Fresnel's double mirror; (c) Fresnel's biprism.

In Lloyd's single-mirror experiment [Figure 3.3(a)], a pinhole source S is placed near a plane mirror as shown. That part of the light that is reflected by the mirror appears to come from a virtual source S'. The field in the region of the screen is then equivalent to that in Young's experiment. In calculating the intensity at a point P the phase change that occurs on reflection must of course be taken into account.

The Fresnel double-mirror arrangement [Figure 3.3(b)] makes use of two mirrors to obtain two mutually coherent virtual sources S' and S'', as shown.

A glass prism is employed to obtain two mutually coherent sources in the Fresnel biprism arrangement [Figure 3.3(c)]. The apex angle of the prism should be nearly 180 degrees in order that the virtual sources are only slightly separated.

Classification of Interference Methods The methods of demonstrating interference mentioned above can be broadly classified as interference by *division of wave front*. In this class we have a single pointlike or linelike source emitting waves in different directions. These waves are eventually brought together by means of mirrors, prisms, and lenses to produce interference fringes. A second class is known as *division of amplitude*. In this case a single beam of light is divided into two or more beams by partial reflection. Here we need not have a point source, since the wave fronts of the reflected and the transmitted beams have a one-to-one correspondence. The Michelson interferometer, described in the following section, illustrates this class of interference method.

3.3 The Michelson Interferometer

Perhaps the best known and most versatile interferometric device is the interferometer developed by Michelson in 1880. The basic design is shown in Figure 3.4. Light from the source S falls on a lightly sil-

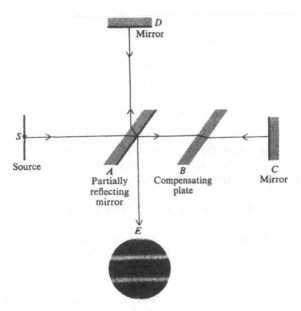

Figure 3.4.   Optical paths in the Michelson interferometer.

vered glass plate A, which divides the beam into two parts. These separated beams are reflected back to A by mirrors C and D as shown. Usually a compensating plate B is inserted in one beam in order that the two optical paths include the same thickness of glass. The compensating plate is necessary when observing fringes with white light.

The interference pattern is observed at E. Here the light appears to come from two virtual source planes H_1 and H_2 as indicated in Figure 3.5. The corresponding virtual point sources S' and S'' in

Figure 3.5. Virtual source planes in the Michelson interferometer.

these planes are mutually coherent. If d is the path difference between the two rays reaching E, that is, the distance between S_1' and S_2'', then from Equations (3.3) and (3.4), the irradiance is proportional to

$$1 + \cos \theta = 1 + \cos kd = 1 + \cos \frac{2\pi d}{\lambda} \qquad (3.10)$$

Now if the mirrors are slightly tilted so that the virtual source planes H_1 and H_2 are not quite parallel, then alternate bright and dark fringes appear across the field of view when the eye is placed at E. These fringes, called *localized fringes*, appear to come from the region of H_1 and H_2. On the other hand if H_1 and H_2 are parallel, then the fringes are seen as circular and appear to come from infinity.

Several localized colored fringes can be observed with white light if H_1 and H_2 intersect at some point in the field of view. In this case the central fringe is dark owing to the fact that one ray is internally reflected in plate A, whereas the other ray is externally reflected in A

and, accordingly, the two rays reaching E are 180 degrees out of phase for $d = 0$.

One of the many uses of the Michelson interferometer is the determination of the index of refraction of gases. An evacuated optical cell is placed in one of the optical paths of the interferometer. The gas whose index is to be measured is then allowed to flow into the cell. This is equivalent to changing the lengths of the optical path and causes the interference fringes to move across the field of view. The number of such fringes gives the effective change in optical path from which the index of refraction of the gas can be calculated.

A modification of the Michelson interferometer known as the Twyman-Green interferometer is shown in Figure 3.6. This inter-

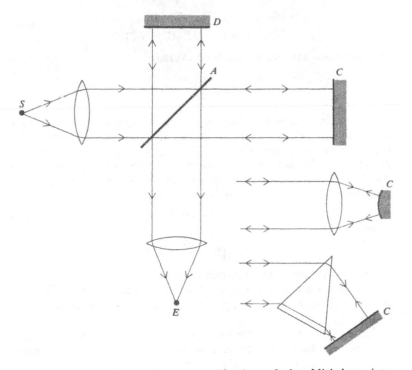

Figure 3.6. The Twyman-Green modification of the Michelson interferometer.

ferometer is used for testing optical elements such as lenses, mirrors, and prisms. Collimated light is used in this case. The optical element to be tested is placed in one of the paths as shown. Imperfections are rendered visible by distortions in the fringe pattern.

A more complete discussion of various types of interferometers is given in References [7] and [40].

3.4 Theory of Partial Coherence. Visibility of Fringes

In the preceding discussions it was assumed that the optical fields were completely coherent, monochromatic, and constant in amplitude. In the actual case of interference of two or more light waves, the amplitudes and phases usually vary with time in a random fashion. The instantaneous light flux at a given point therefore fluctuates rapidly. It would seem more meaningful then to define the irradiance as a time average. In the case of two fields E_1 and E_2 the irradiance I can be expressed accordingly as

$$I = \langle \mathbf{E} \cdot \mathbf{E}^* \rangle = \langle (\mathbf{E}_1 + \mathbf{E}_2) \cdot (\mathbf{E}_1^* + \mathbf{E}_2^*) \rangle$$
$$= \langle |\mathbf{E}_1|^2 + |\mathbf{E}_2|^2 + 2Re(\mathbf{E}_1 \cdot \mathbf{E}_2^*) \rangle \qquad (3.11)$$

The sharp brackets denote the time average:

$$\langle f \rangle = \lim_{T \to \infty} \frac{1}{T} \int_0^T f(t) \, dt \qquad (3.12)$$

In the discussions to follow, it will be assumed that all quantities are *stationary*. By stationary is meant that the time average is independent of the choice of the origin of time. Also, for convenience, the optical fields will be assumed to have the same polarization so that their vectorial nature can be ignored. With these simplifying assumptions, Equation (3.11) can be written

$$I = I_1 + I_2 + 2Re\langle E_1 E_2^* \rangle \qquad (3.13)$$

where

$$I_1 = \langle |E_1|^2 \rangle \qquad I_2 = \langle |E_2|^2 \rangle \qquad (3.14)$$

In the usual interference experiment the two fields E_1 and E_2 originate from some common source. They differ because of a difference in their optical paths. A simplified schematic diagram is shown in Figure 3.7.

Let us call t the time for one light signal to traverse path 1 and $t + \tau$ the time for the other signal to traverse path 2. Then the interference term in Equation (3.13) may be written as

$$2Re\Gamma_{12}(\tau)$$

where

$$\Gamma_{12}(\tau) = \langle E_1(t)E_2^*(t + \tau) \rangle \qquad (3.15)$$

The function $\Gamma_{12}(\tau)$ is called the *mutual coherence function* or the *correlation function* of two fields E_1 and E_2.

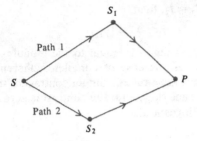

Figure 3.7. Generalized light paths in an interference experiment.

The function

$$\Gamma_{11}(\tau) = \langle E_1(t)E_1^*(t + \tau)\rangle$$

is known as the *autocorrelation function* or the *self-coherance function*. From the definition, we see that $\Gamma_{11}(0) = I_1$ and $\Gamma_{22}(0) = I_2$.

It is sometimes convenient to use a normalized correlation function called the *degree of partial coherence:*

$$\gamma_{12}(\tau) = \frac{\Gamma_{12}(\tau)}{\sqrt{\Gamma_{11}(0)\Gamma_{22}(0)}} = \frac{\Gamma_{12}(\tau)}{\sqrt{I_1 I_2}} \qquad (3.16)$$

The irradiance is then expressed as follows:

$$I = I_1 + I_2 + 2\sqrt{I_1 I_2}\, Re\, \gamma_{12}(\tau) \qquad (3.17)$$

The function $\gamma_{12}(\tau)$ is, in general, a complex periodic function of τ. Thus an interference pattern results if $|\gamma_{12}(\tau)|$ has a value other than zero. In terms of $|\gamma_{12}(\tau)|$ we have the following types of coherence:

$$|\gamma_{12}| = 1 \qquad \text{(complete coherence)}$$
$$0 < |\gamma_{12}| < 1 \qquad \text{(partial coherence)}$$
$$|\gamma_{12}| = 0 \qquad \text{(complete incoherence)}$$

In a pattern of interference fringes, the intensity varies between two limits I_{max} and I_{min}. From Equation (3.17) we see that these are given by

$$I_{max} = I_1 + I_2 + 2\sqrt{I_1 I_2}\,|\gamma_{12}| \qquad I_{min} = I_1 + I_2 - 2\sqrt{I_1 I_2}\,|\gamma_{12}| \quad (3.18)$$

The *fringe visibility* $\mathscr{V}$ is defined as the ratio

$$\mathscr{V} = \frac{I_{max} - I_{min}}{I_{max} + I_{min}} \qquad (3.19)$$

It follows that

$$\mathscr{V} = \frac{2\sqrt{I_1 I_2}\,|\gamma_{12}|}{I_1 + I_2} \qquad (3.20)$$

In particular, if $I_1 = I_2$, then

$$\mathscr{V} = |\gamma_{12}| \tag{3.21}$$

that is, the fringe visibility is equal to the modulus of the degree of partial coherence. In the case of complete coherence ($|\gamma_{12}| = 1$) the interference fringes have the maximum contrast of unity, whereas for complete incoherence ($|\gamma_{12}| = 0$) the contrast is zero; that is, there are no interference fringes at all.

3.5　Coherence Time and Coherence Length

In order to see how the degree of partial coherence is related to the characteristics of the source, let us consider the case of a hypothetical "quasimonochromatic" source having the following property: The oscillation and the subsequent field vary sinusoidally for a certain time τ_0 and then change phase abruptly. This sequence keeps repeating indefinitely. A graph is shown in Figure 3.8. We shall call τ_0 the

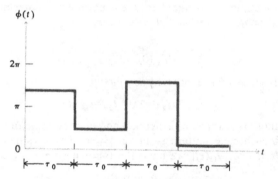

Figure 3.8.　Graph of the phase $\phi(t)$ of a quasimonochromatic source.

coherence time. The phase change that occurs after each coherence time is considered to be randomly distributed between 0 and 2π.

　　The time dependence of this quasimonochromatic field can be expressed as

$$E(t) = E_0 e^{-i\omega t} e^{i\phi(t)} \tag{3.22}$$

where the phase angle $\phi(t)$ is a random step function, illustrated in Figure 3.8. One can regard the above kind of a field as an approximation to that of a radiating atom, the abrupt changes of phase being the result of collisions.

　　Suppose a beam of light, whose field is represented by Equation (3.22), is divided into two beams that are subsequently brought together to produce interference. The degree of partial coherence can be evaluated as follows: It is assumed that

$$|E_1| = |E_2| = |E|$$

Then, since we are concerned here with *self-coherence*, we delete the subscripts and write

$$\gamma(\tau) = \frac{\langle E(t)\, E^*(t + \tau)\rangle}{\langle |E|^2\rangle} \tag{3.23}$$

From (3.22) we have

$$\gamma(\tau) = \langle e^{i\omega\tau}\, e^{i[\phi(t)-\phi(t+\tau)]}\rangle$$

$$= e^{i\omega\tau}\lim_{T\to\infty}\frac{1}{T}\int_0^T e^{i[\phi(t)-\phi(t+\tau)]}\, dt \tag{3.24}$$

Consider the quantity $\phi(t) - \phi(t + \tau)$. This is plotted in Figure 3.9.

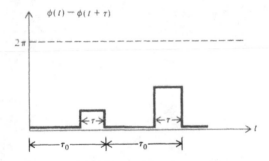

Figure 3.9. Graph of the phase difference $\phi(t) - \phi(t + \tau)$.

Now for the first coherence time interval, $0 < t < \tau_0$, we observe that $\phi(t) - \phi(t + \tau) = 0$ for $0 < t < \tau_0 - \tau$. On the other hand for $\tau_0 - \tau < t < \tau_0$ it assumes some random value between 0 and 2π. The same is true for each succeeding coherence time interval.

The integral in Equation (3.24) is easily evaluated as follows: For the first interval we have

$$\frac{1}{\tau_0}\int_0^{\tau_0} e^{i[\phi(t)-\phi(t+\tau)]}\, dt = \frac{1}{\tau_0}\int_0^{\tau_0-\tau} dt + \frac{1}{\tau_0}\int_{\tau_0-\tau}^{\tau_0} e^{i\Delta}\, dt \tag{3.25}$$

$$= \frac{\tau_0 - \tau}{\tau_0} + \frac{\tau}{\tau_0}\, e^{i\Delta}$$

where Δ is the random phase difference.

The same result is obtained for all subsequent intervals, except that Δ is different for each interval. Since Δ is random, the terms involving $e^{i\Delta}$ will average to zero. The other term, $(\tau_0 - \tau)/\tau_0$, is the same for all intervals, hence it is equal to average value of the integral in question. Of course if $\tau > \tau_0$, then the phase difference $\phi(t) - \phi(t + \tau)$ is always random and, consequently, the whole integral averages to zero.

From the above result we find that the normalized autocorrelation function for a quasimonochromatic source is given by

$$\gamma(\tau) = \left(1 - \frac{\tau}{\tau_0}\right) e^{i\omega\tau} \qquad \tau < \tau_0$$
$$= 0 \qquad \tau \geq \tau_0 \tag{3.26}$$

$$|\gamma(\tau)| = 1 - \frac{\tau}{\tau_0} \qquad \tau < \tau_0$$
$$= 0 \qquad \tau \geq \tau_0 \tag{3.27}$$

A graph of $|\gamma|$ is shown in Figure 3.10. We found in the previous section that this quantity is

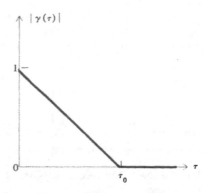

Figure 3.10.   Graph of the coherence of a quasimonochromatic source.

equal to the fringe visibility $\mathscr{V}$ for the case of equal amplitudes in a two-beam interference arrangement. Evidently the fringe visibility drops to zero if τ exceeds the coherence time τ_0. This means that the path difference between the two beams must not exceed the value

$$c\tau_0 = l_c$$

in order to obtain interference fringes. The quantity l_c is called the *coherence length*. It is essentially the length of an uninterrupted wave train.

In the actual case of radiating atoms, the time between collisions is not constant but varies randomly from one collision to the next. Consequently, the wave trains vary in length in a similar random fashion. In this more realistic situation we can define the coherence time as the *average* value of the individual coherence times and similarly for the coherence length. The actual mathematical form of the degree of coherence and of the fringe visibility will then depend on the precise statistical distribution of the lengths of the wave trains. In any case the fringe visibility will be large (of the order of unity) for path differences that are small compared to the average coherence length. Conversely, the fringe visibility will become small and approach zero as the path difference becomes larger than the average coherence length.

For further reading on the subject of coherence, References [1] and [5] may be consulted.

3.6 Spectral Resolution of a Finite Wave Train. Coherence and Line Width

In practice no source of light is ever strictly monochromatic. Even in the best so-called monochromatic sources there is always some finite spread of frequency centered about some mean frequency. We now proceed to investigate the relationship between the frequency spread, or line width, and the coherence of a light source. To do this we make use of the Fourier integral theorem [27].

According to the theorem, stated here without proof, a function $f(t)$ can be expressed as an integral over the variable ω in the following way:

$$f(t) = \frac{1}{\sqrt{2\pi}} \int_{-\infty}^{+\infty} g(\omega) e^{-i\omega t} \, d\omega$$
$$g(\omega) = \frac{1}{\sqrt{2\pi}} \int_{-\infty}^{+\infty} f(t) e^{i\omega t} \, dt$$

(3.28)

The functions $f(t)$ and $g(\omega)$ are called *Fourier transforms* of each other and are said to constitute a Fourier transform pair. In our present application the variables t and ω are time and frequency, respectively. The function $g(\omega)$ then constitutes a frequency resolution of the time dependent function $f(t)$ or stated in another way, $g(\omega)$ represents the function in the frequency domain.

Let us consider now the particular case in which the function $f(t)$ represents a single wave train of finite duration τ_0. The time variation of this wave train is given by the function

$$f(t) = e^{-i\omega_0 t} \quad \text{for } -\frac{\tau_0}{2} < t < \frac{\tau_0}{2}$$

(3.29)

$$f(t) = 0 \quad \text{otherwise}$$

Taking the Fourier transform, we have

$$g(\omega) = \frac{1}{\sqrt{2\pi}} \int_{-\tau_0/2}^{+\tau_0/2} e^{i(\omega - \omega_0)t} \, dt$$

$$= \sqrt{\frac{2}{\pi}} \, \frac{\sin \, [(\omega - \omega_0)\tau_0/2]}{\omega - \omega_0} \qquad (3.30)$$

A plot of the real part of the function $f(t)$ is shown in Figure 3.11.

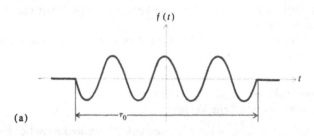

(a)

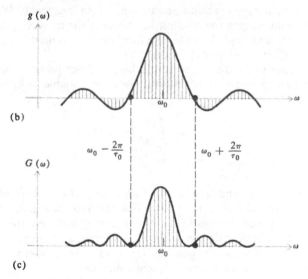

(b)

(c)

Figure 3.11.   (a) A finite wave train, (b) its Fourier transform, and (c) the power spectrum.

Also plotted is a graph of the *power spectrum:*

$$G(\omega) = |g(\omega)|^2$$

This function, in the case of a finite wave train, is given by

$$G(\omega) = |g(\omega)|^2 = \frac{2 \, \sin^2 \, [(\omega - \omega_0)\tau_0/2]}{\pi(\omega - \omega_0)^2} \qquad (3.31)$$

We see that the spectral distribution is maximum for $\omega = \omega_0$ and drops to zero for $\omega = \omega_0 \pm 2\pi/\tau_0$. Secondary maxima and minima also occur as shown in the diagram. Most of the energy is contained in the region between the first two minima on either side of the central maximum at ω_0. The "width" $\Delta\omega$ of the frequency distribution is therefore given by

$$\Delta\omega = \frac{2\pi}{\tau_0} \qquad (3.32)$$

or

$$\Delta\nu = \frac{1}{\tau_0} \qquad (3.33)$$

Now if we have a sequence of wave trains, each lasting for a time τ_0 but occurring at random times, then the power spectrum is exactly the same as that of the single pulse given above. On the other hand, if the pulses are not all of the same duration, that is, if τ_0 varies from pulse to pulse, then we can think of an average time $\langle\tau_0\rangle$. The precise form of the spectral distribution is different from that of the single pulse, but the width of the corresponding frequency spectrum is approximately $\langle\tau_0\rangle^{-1}$. Suppose we now take the reverse reasoning, namely, that if a spectral source has a line width $\Delta\nu$, then the corresponding coherence time $\langle\tau_0\rangle$ is given by

$$\langle\tau_0\rangle = \frac{1}{\Delta\nu} \qquad (3.34)$$

and the coherence length l_c is

$$l_c = c\langle\tau_0\rangle = \frac{c}{\Delta\nu} \qquad (3.35)$$

We can also express the coherence length in terms of wavelength. Using the fact that $\Delta\nu/\nu = |\Delta\lambda|/\lambda$, we obtain

$$l_c = \frac{\lambda^2}{\Delta\lambda} \qquad (3.36)$$

where $\Delta\lambda$ is the width of the spectrum line on the wavelength scale.

As a specific example, ordinary spectral sources, such as discharge tubes, have line widths of the order of an angstrom in the visible region of the spectrum, ~ 5000 Å. The corresponding coherence length, from Equation (3.36), is of the order of 5000 wavelengths or about 2 mm. In an interference experiment the fringe visibility would become vanishingly small for path differences much larger than this distance.

In the case of an interference experiment with white light in which the eye is used to detect the interference fringes, we must consider the spectral sensitivity of the eye. This is maximum at about 5500 Å, and falls to zero at approximately 4000 Å and 7000 Å.

Therefore, as far as the eye is concerned, the spectral "width" of a white-light source is about 1500 Å, and the corresponding coherence length is about 3 or 4 wavelengths. This is about the number of fringes that can be seen on either side of the zero fringe in the Michelson interferometer when using a source of white light such as a tungsten lamp.

At the other extreme, the line width of a gas laser may be as narrow as 10^3 Hz or less. This corresponds to a coherence length of $\nu/\Delta\nu \approx 10^{14}/10^3 = 10^{11}$ wavelengths, which is the order of 50 km. Not only can interference effects be produced over very long distances with lasers, but interference fringes can be produced by using *two different* lasers as sources. However, if two lasers are used, the fringe pattern is not steady but fluctuates in a random fashion. A given fringe pattern persists for a time interval of the order of the coherence time of the laser sources. Typically this is about 10^{-3} s.

3.7 Spatial Coherence

In the previous section we treated the problem of coherence between two fields arriving at the same point in space over different optical paths. We now wish to discuss the more general problem of coherence between two fields at different points in space. This will be of importance in studying the coherence of the radiation fields of extended sources.

Suppose, first, that we have a single quasimonochromatic point source S (Figure 3.12). Three receiving points, P_1, P_2, P_3, are located

Figure 3.12.   Diagram to illustrate lateral and longitudinal coherence.

as shown. The corresponding fields are E_1, E_2, and E_3, respectively. The two points P_1 and P_3 lie in the same direction from the source. They differ only in their distances from S. Accordingly, the coherence between the fields E_1 and E_3 measures the *longitudinal spatial coherence* of the field. On the other hand the receiving point P_2 is located at the same distance from S as P_1. In this case the coherence between E_1 and E_2 measures the *transverse spatial coherence* of the field.

It is evident that the longitudinal coherence will merely depend on how large r_{13} is in comparison with the coherence length of the source or, equivalently, on the value of $t_{13} = r_{13}/c$ compared to the coherence time τ_0. For whatever $E_1(t)$ is, E_3 will vary with t in the same way but at a time t_{13} later. If $t_{13} \ll \tau_0$, there will be a high coherence between E_1 and E_3, whereas if $t_{13} \gg \tau_0$, there will be little or no coherence.

Regarding the transverse coherence, if S is a true point source, then the time dependence of the two fields E_1 and E_2 will be precisely the same; that is, they will be completely mutually coherent. Partial coherence between E_1 and E_2 will occur if the source has spatial extension rather than being a point. We shall now proceed to find how the transverse coherence of the field is related to the size of the source.

Since an extended source can be considered to be made up of many independent point sources, it will be convenient to study the case of two point sources before discussing the more general case of finite extended sources. The geometry is shown in Figure 3.13. The

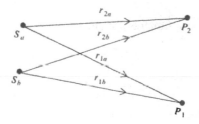

Figure 3.13. Geometry for analyzing the lateral coherence from two sources.

two quasimonochromatic point sources S_a and S_b are considered to be identical with one another, except, of course, that their phases vary randomly and independently. In other words they are mutually incoherent.

We have

$$E_1 = E_{1a} + E_{1b}$$
$$E_2 = E_{2a} + E_{2b}$$

where E_{1a} is the contribution to the field at P_1 from the source S_a, and similarly for E_{1b}, and so forth.

The normalized correlation function for the two receiving points is then given by

$$\begin{aligned}
\gamma_{12}(\tau) &= \frac{\langle E_1(t)E_2^*(t+\tau)\rangle}{\sqrt{I_1 I_2}} \\
&= \frac{\langle [E_{1a}(t) + E_{1b}(t)][E_{2a}^*(t+\tau) + E_{2b}^*(t+\tau)]\rangle}{\sqrt{I_1 I_2}} \\
&= \frac{\langle E_{1a}(t)E_{2a}^*(t+\tau)\rangle}{\sqrt{I_1 I_2}} + \frac{\langle E_{1b}(t)E_{2b}^*(t+\tau)\rangle}{\sqrt{I_1 I_2}}
\end{aligned} \tag{3.37}$$

In the second step we have used the fact that S_a and S_b are mutually incoherent, so the cross terms $\langle E_{1a} E_{2b}^{*} \rangle$ and $\langle E_{1b} E_{2a}^{*} \rangle$ both vanish.

If it is assumed that the fields are both of the type given by Equation (3.22) in Section 3.5, then the two time averages in the above equation can be evaluated in exactly the same way that was done in arriving at Equation (3.25). It is necessary, however, to take into account the different times for the optical fields to travel from their respective sources to the receiving points. If this is done, the result can be expressed as follows:

$$\gamma_{12}(\tau) = \tfrac{1}{2}\gamma(\tau_a) + \tfrac{1}{2}\gamma(\tau_b) \tag{3.38}$$

where

$$\gamma(\tau) = e^{i\omega\tau}\left(1 - \frac{\tau}{\tau_0}\right)$$

is the autocorrelation function of either source, and

$$\tau_a = \frac{r_{2a} - r_{1a}}{c} + \tau$$

$$\tau_b = \frac{r_{2b} - r_{1b}}{c} + \tau$$

After some algebra, we find that

$$|\gamma_{12}(\tau)|^2 \approx \left(\frac{1 + \cos\,[\omega(\tau_b - \tau_a)]}{2}\right)\left(1 - \frac{\tau_a}{\tau_0}\right)\left(1 - \frac{\tau_b}{\tau_0}\right) \tag{3.39}$$

In arriving at the above result it has been assumed that $\tau_a - \tau_b$ is small in comparison with τ_a and with τ_b.

The analysis above shows that the mutual coherence between the fields at the two receiving points depends not only on the self-coherence time τ_0 of the sources, but also on the quantity $\tau_b - \tau_a$ in a periodic manner as expressed by the cosine term in Equation (3.39). In other words the mutual coherence between a given fixed receiving point and any other point being illuminated by the two mutually incoherent sources exhibits a *periodic spatial dependence* that is somewhat like an interference pattern although the total illumination is quite uniform.

Suppose, for example, we take the point P_1 to be symmetrically located with respect to the two sources so that $r_{1a} = r_{1b}$ in Figure 3.13. We then have $\tau_b - \tau_a = (r_{2b} - r_{1b})/c$, or approximately

$$\tau_b - \tau_a \approx \frac{sl}{2cr} \tag{3.40}$$

in which s is the distance between the two sources, l is the distance between the two receiving points, and r is the mean distance from the sources to the receiving points. The approximate result is obtained on the assumption that r is very large compared to both s and l. The geometry is similar to that of the Young interference experiment. In Figure 3.14 the variation of $|\gamma_{12}|$ is illustrated as a curve with

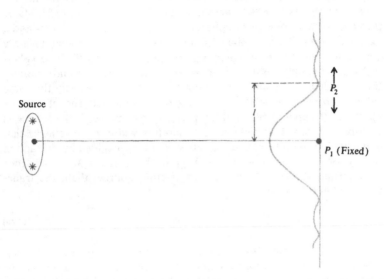

Figure 3.14. Diagram illustrating the lateral coherence of an extended source.

maxima and minima. The mutual coherence is greatest at the center where P_1 and P_2 coincide. The coherence drops to zero on either side of the central line at a distance l_t such that $\cos\left[\omega(\tau_b - \tau_a)\right] = -1$; that is,

$$\omega(\tau_b - \tau_a) = \frac{\omega s l_t}{2cr} = \pi \qquad (3.41)$$

Since $\omega = 2\pi c/\lambda$, we find

$$l_t = \frac{r\lambda}{s} \qquad (3.42)$$

In terms of the angular separation θ_s between the two sources, as seen at the receiving point P_1, we have $\theta_s \cong s/r$, and so

$$l_t = \frac{\lambda}{\theta_s} \qquad (3.43)$$

This is approximately the width of the region of high mutual coherence between P_1 and P_2. We shall call it the *transverse coherence width*.

Extended Sources. The Van Cittert-Zernike Theorem   In the case of an extended primary source, such as the surface of the sun or an incandescent lamp, the mathematical problem of calculating the mutual coherence between two points is fairly difficult. A general theorem, known as the *Van Cittert-Zernike theorem,* is very useful in this case. The theorem, which we shall not attempt to prove [5], states that the complex degree of coherence between a fixed point P_1 and a variable point P_2 in a plane illuminated by an extended primary source is equal to the complex amplitude produced at P_2 by a spherical wave passing through an aperture of the same size and shape as the extended source and converging to P_1. This is precisely the same calculation as is done in computing diffraction patterns of various apertures (treated in the next chapter). For example, if the source is circular in shape, then the lateral coherence width is the same as that given by Equation (3.43) with a numerical factor (given by diffraction theory) of 1.22, which, as we shall see in the next chapter, is the root of a certain Bessel function. For a circular source, then, the transverse coherence width is

$$l_t = \frac{1.22\lambda}{\theta_s} \tag{3.44}$$

According to the above result, if one were to perform a Young-type interference experiment using a pinhole source and a double-slit aperture, then the distance between the slits would have to be less than the transverse coherence width in order to observe distinct interference fringes. As a numerical example, suppose a pinhole of 1-mm diameter is the source, and the wavelength is, say, 600 nm. Then at a distance of 1 m from the source the transverse coherence width, from Equation (3.44), will turn out to be about 0.7 mm.

Measurement of Stellar Diameters   Due to their enormous distances, the angular diameters of stars are extremely small, of the order of

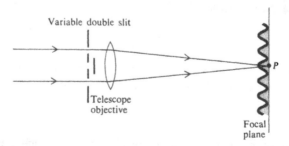

Figure 3.15.   Method for producing interference fringes from a distant source.

hundredths of an arc second. Hence the transverse coherence width of the light from a star, as viewed from the earth, is of the order of several meters. A method of determining the angular diameter of a distant source of light, such as a star, is to use a double-slit interference arrangement in which the slit separation can be varied, as shown in Figure 3.15. The transverse coherence width is first determined. It is merely the slit separation that results in the disappearance of interference fringes. The angular diameter of the source is then given by Equation (3.44).

Michelson was the first to determine stellar diameters by interferometry. He employed mirrors to increase the distance between the slits (Figure 3.16). One of the largest stars measured, Betelgeuse, was

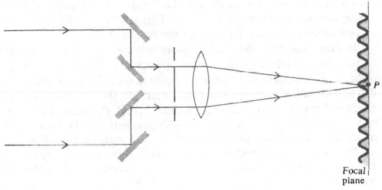

Figure 3.16. Michelson's stellar interferometer.

found to have an angular diameter of 0.047 arc second. From the known distance, this corresponds to a linear diameter of about 280 times that of the sun.

3.8 Intensity Interferometry

A method of interferometry based on intensity correlations between two points has been devised by Hanbury-Brown and Twiss. This method, known as *intensity interferometry,* makes it possible to determine much smaller stellar angular diameters than those measureable by Michelson's method.

The essential features of the Hanbury-Brown–Twiss arrangement are shown in Figure 3.17. There are two mirrors that need not be of high optical quality. (Searchlight mirrors were actually used.) The light is focused onto photocells the outputs of which are proportional to the instantaneous intensities $|E_1|^2$ and $|E_2|^2$ at the two mirrors. The

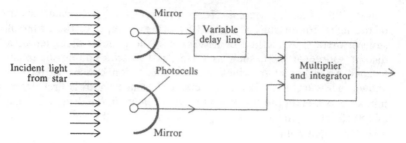

Figure 3.17.   The Hanbury-Brown–Twiss intensity interferometer.

signals from the photocells are fed to a delay line and an electronic multiplier and integrator as shown. The output is proportional to the average of the product $\langle |E_1|^2 \; |E_2|^2 \rangle$. This quantity is known as the *second-order coherence function* of the two fields.

It turns out that the second-order coherence function exhibits an interference effect similar to the interference effect shown by the ordinary (first-order) function already discussed. In particular, for a distant extended source, a measurement of the second-order coherence between two receiving points P_1 and P_2 yields the lateral coherence width and, hence, the angular diameter of the source. The main advantage of the method of intensity interferometry is that high-quality optical components and rigid mountings are not necessary. For more information on this subject, the reader should consult the References [10] [15] listed at the end of the book.

3.9   Fourier Transform Spectroscopy

Suppose that a beam of light is divided into two mutually coherent beams, as in the Michelson interferometer, and these beams are reunited after traversing different optical paths (Figure 3.18). If the light is not monochromatic but has a spectral composition given by the function $G(\omega)$, then the intensity at P varies in a manner that depends on the particular spectrum. By recording the intensity as a function of the path difference, the power spectrum $G(\omega)$ can be deduced. This method of obtaining a spectrum is called *Fourier transform spectroscopy*.

For the present application it is convenient to represent the spectral distribution in terms of the wavenumber k, rather than the angular frequency ω. Since ω and k are proportional to one another ($\omega = ck$ in vacuum), we can just as well use $G(k)$ as $G(\omega)$. Now, referring to Equation (3.3), which gives the intensity at P for monochromatic

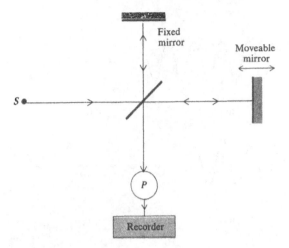

Figure 3.18. Arrangement for Fourier transform spectroscopy.

light, we see that the intensity for nonmonochromatic light will be given by a summation over the complete spectrum, namely,

$$I(x) = \int_0^\infty (1 + \cos kx)\, G(k)\, dk$$

$$= \int_0^\infty G(k)\, dk + \int_0^\infty G(k)\, \frac{e^{ikx} + e^{-ikx}}{2}\, dk$$

$$= \tfrac{1}{2}I(0) + \tfrac{1}{2} \int_{-\infty}^\infty G(k) e^{ikx}\, dk$$

or

$$W(x) = 2I(x) - I(0) = \int_{-\infty}^\infty e^{ikx} G(k)\, dk\, \frac{1}{\sqrt{2\pi}} \qquad \text{(3.45)}$$

where $I(0)$ is the intensity for zero path difference. Therefore $W(x)$ and $G(k)$ constitute a Fourier transform pair. Accordingly, we can write

$$G(k) = \frac{1}{\sqrt{2\pi}} \int_{-\infty}^\infty W(x) e^{-ikx}\, dx \qquad \text{(3.46)}$$

that is, the power spectrum $G(k)$ is the Fourier transform of the intensity function $W(x) = 2I(x) - I(0)$.

The above technique of spectrum analysis is particularly useful for analyzing the infrared absorption of gases where the spectra are extremely complicated. A second advantage is the efficient utilization of the available light. This makes the Fourier transform method invaluable for the study of very weak sources.

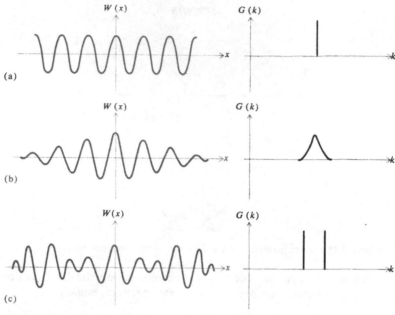

Figure 3.19.   Intensity functions and their spectra. (a) A single monochromatic line; (b) a single broad line; (c) two narrow lines. (Spectra for positive values of k only are shown.)

The actual calculation of the Fourier transform of the intensity function is often done by means of high-speed electronic computers. Examples of some intensity functions and the corresponding spectra are shown in Figure 3.19.

PROBLEMS

3.1   Calculate the interference pattern that would be obtained if three identical slits instead of two were used in Young's experiment. (Assume equal spacing of the slits.) Make a rough plot.

3.2   In a two-slit interference experiment of the Young type, the aperture-to-screen distance is 2 m and the wavelength is 600 nm. If it is desired to have a fringe spacing of 1 mm, what is the required slit separation?

3.3   In the experiment of Problem 3.2 a thin plate of glass ($n = 1.5$) of thickness 0.05 mm is placed over one of the slits. What is the resulting lateral fringe displacement at the screen?

3.4   In Lloyd's single-mirror interference experiment, the angle of incidence is $90°-\alpha$, where α is very small. Calculate the fringe spacing in terms of α, the wavelength λ, the slit separation h, and the aperture-to-screen distance x. Assume that the mirror is halfway between the aperture and the screen.

3.5   A Fresnel biprism has an apex angle $180°-\alpha$, where α is very small. The index of refraction is n, the distance from the source to the biprism is D, and the distance from the biprism to the screen is D'. Find the fringe spacing. Use appropriate approximations. Assume $\lambda = 600$ nm.

3.6   A Michelson interferometer can be used to determine the index of refraction of a gas. The gas is made to flow into an evacuated glass cell of length l placed in one arm of the interferometer. The interference fringes are counted as they move across the view aperture when the gas flows into the cell. Show that the effective optical path difference of the light beam for the full cell versus the evacuated cell is $2l(n-1)$, where n is the index of refraction of the gas, and hence that a number $N = 4l(n-1)/\lambda$ fringes move across the field of view as the cell is filled. How many fringes would be counted if the gas were air ($n = 1.0003$) for a 10-cm cell using yellow sodium light $\lambda = 590$ nm?

3.7   A filter is used to obtain approximately monochromatic light from a white source. If the pass band of the filter is 10 nm, what is the coherence length and coherence time of the filtered light? The mean wavelength is 600 nm.

3.8   What is the line width in hertz and in nanometers of the light from a helium-neon laser whose coherence length is 5 km? The wavelength is 633 nm.

3.9   What is the transverse coherence width of sunlight? The apparent angular diameter of the sun is 0.5 degrees and the mean effective wavelength is 600 nm.

3.10 A pinhole of 0.5 mm diameter is used as a source for a Young double-slit interference experiment. A sodium lamp ($\lambda = 590$ nm) is used. If the distance from the source to the double-slit aperture is 0.5 m, what is the maximum slit spacing such that interference fringes are just observable?

3.11 A tungsten-filament lamp having a straight filament that is 0.1 mm in diameter is used as a source for an interference experiment. At what distance from the lamp must the aperture be placed in order that the transverse coherence width is at least 1 mm? If a double-slit aperture is used, why should the slits be oriented parallel to the lamp filament?

3.12 Calculate the power spectrum of a damped wave train:

$$f(t) = A \exp(-at - i\omega_0 t), \quad t \geq 0$$
$$f(t) = 0, \qquad\qquad\qquad\quad t < 0$$

3.13 Show that the power spectrum of a Gaussian pulse

$$f(t) = A \exp(-at^2 - i\omega_0 t)$$

is also a Gaussian function centered at the frequency ω_0.

CHAPTER 4
Multiple-Beam Interference

4.1 Interference with Multiple Beams

In our study of interference thus far we have been concerned only with interference between two beams. We now proceed to treat the more general case of multiple-beam interference.

The most common method of producing a large number of mutually coherent beams is by division of amplitude. The division occurs by multiple reflection between two parallel, partially reflecting surfaces. These surfaces might be semitransparent mirrors, or merely the two sides of a film or slab of transparent material, and so forth. The situation is illustrated in Figure 4.1. Here, for simplicity, the reflecting surfaces are considered to be two thin, identical semireflecting mirrors. The primary beam is partially reflected and partially transmitted at the first surface. The transmitted part is subsequently reflected back and forth between the two surfaces as shown.

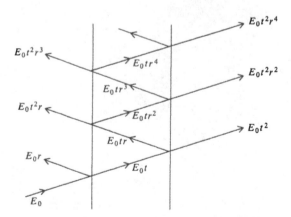

Figure 4.1. Paths of light rays in multiple reflection between two parallel mirrors. (For simplicity the mirrors are considered to be infinitely thin.)

Let r be the coefficient of reflection and t the transmission coefficient. Then, provided there is no absorption in the medium between the reflecting surfaces, the amplitudes of the successive internally reflected rays are $E_0 t$, $E_0 tr$, $E_0 tr^2$, . . . , as indicated in the figure. Here E_0 is the amplitude of the primary ray. Consequently, the sequence $E_0 t^2$, $E_0 t^2 r^2$, $E_0 t^2 r^4$, . . . represents the amplitudes of the successive transmitted rays. Now it is easily shown that the geometric path difference between any two successive transmitted rays is $2d \cos \theta$, where d is the separation between the two reflecting surfaces and θ is the angle between any internally reflected ray and the surface normal, as indicated in Figure 4.2. The corresponding phase dif-

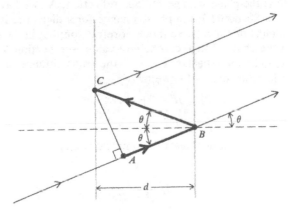

Figure 4.2. Diagram showing the path difference between two successive rays.

ference between any two successive rays is then given by

$$\delta = 2kd \cos \theta = \frac{4\pi}{\lambda} d \cos \theta \qquad (4.1)$$

where λ is the wavelength in the medium. In terms of the vacuum wavelength λ_0 we have

$$\delta = \frac{4\pi}{\lambda_0} nd \cos \theta \qquad (4.2)$$

in which n is the index of refraction of the medium between the reflecting surfaces. Taking this phase difference into account as a factor $e^{i\delta}$ and summing the amplitudes of the transmitted rays, we obtain

$$E_T = E_0 t^2 + E_0 t^2 r^2 e^{i\delta} + E_0 t^2 r^4 e^{2i\delta} + \cdots$$

This is a geometric series with ratio $r^2 e^{i\delta}$, hence

$$E_T = \frac{E_0 t^2}{1 - r^2 e^{i\delta}}$$

The intensity $I_T = |E_T|^2$ of the transmitted light is thus given by

$$I_T = I_0 \frac{|t|^4}{|1 - r^2 e^{i\delta}|^2} \tag{4.3}$$

where $I_0 = |E_0|^2$ is the intensity of the incident beam.

Now a phase change may occur on reflection, hence r is, in general, a complex number. We can express it, accordingly, as

$$r = |r| e^{i\delta_r/2} \tag{4.4}$$

where $\delta_r/2$ is the phase change for one reflection. As we have shown previously in Section 2.7, the phase change for a dielectric is either 0 or π, depending on the relative index of refraction; but in the case of a metal film the phase change can be any value (see Section 6.6).

If R denotes the reflectance and T the transmittance of one surface, then in terms of r and t, we have

$$R = |r|^2 = rr^*$$
$$T = |t|^2 = tt^* \tag{4.5}$$

in which the asterisk denotes the complex conjugate. Equation (4.3) can then be written

$$I_T = I_0 \frac{T^2}{|1 - R e^{i\Delta}|^2} \tag{4.6}$$

Here we have introduced the abbreviation Δ for the total phase difference between two successive beams

$$\Delta = \delta + \delta_r \tag{4.7}$$

Now

$$|1 - R e^{i\Delta}|^2 = (1 - R e^{i\Delta})(1 - R e^{-i\Delta}) = 1 - R(e^{i\Delta} + e^{-i\Delta}) + R^2$$

$$= 1 - 2R \cos \Delta + R^2 = (1 - R)^2 \left[1 + \frac{4R}{(1 - R)^2} \sin^2 \frac{\Delta}{2} \right]$$

Hence the formula for the intensity can be expressed as

$$I_T = I_0 \frac{T^2}{(1 - R)^2} \frac{1}{1 + F \sin^2 \frac{\Delta}{2}} \tag{4.8}$$

The last term

$$\frac{1}{1 + F \sin^2 \frac{\Delta}{2}}$$

is known as the *Airy function*. The quantity

$$F = \frac{4R}{(1 - R)^2} \tag{4.9}$$

is called the *coefficient of finesse*. It is a measure of the sharpness of the interference fringes.

The general behavior of the Airy function is shown in Figure 4.3.

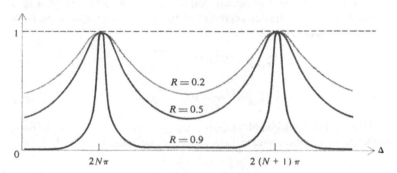

Figure 4.3. Graphs of the Airy function giving the intensity distribution of fringes in multiple-beam interference.

Curves are drawn for various values of the reflectance R and illustrate the intensity distribution of the fringes in multiple-beam interference.

If the argument $\Delta/2$ of the sine term in the Airy function $[1 + F \sin^2(\Delta/2)]^{-1}$ is an integral multiple of π, then the function is equal to its maximum value of unity for any value of F. If the reflectance R is very small, so that F is small, the interference fringes are broad and indistinct; whereas if R is close to unity, which makes F large, then the fringes are very sharp.

The condition for a fringe maximum $\Delta/2 = N\pi$, in which N is an integer, is equivalent to

$$2N\pi = \frac{4\pi}{\lambda_0} nd \cos \theta + \delta_r \tag{4.10}$$

from Equations (4.2) and (4.7). The integer N is known as the *order of interference*. It is equal to the equivalent path difference, measured in wavelengths, between two successive beams.

In the above discussion it has been assumed that the two reflecting surfaces were identical. In the more general case in which the coefficients of reflection are different, namely, $r_1 = |r_1|e^{i\delta_1}$ and $r_2 = |r_2|e^{i\delta_2}$, then it is easy to show that the preceding formulas are correct if we set

$$T = |t_1||t_2| = \sqrt{T_1 T_2}$$
$$R = |r_1||r_2| = \sqrt{R_1 R_2}$$

(4.11)

$$\delta_r = \frac{\delta_1 + \delta_2}{2}$$

(4.12)

The maximum and minimum values of I_T are found by setting Δ equal to 0 and π, respectively. We then find the corresponding fractional transmission to be

$$\mathcal{T}_{max} = \frac{I_{T(max)}}{I_0} = \frac{T^2}{(1-R)^2}$$

(4.13)

$$\mathcal{T}_{min} = \frac{I_{T(min)}}{I_0} = \frac{T^2}{(1+R)^2}$$

(4.14)

If A is the fraction of incident energy that is absorbed at each reflection, then by conservation of energy, we must have

$$A + R + T = 1$$

If there were no absorption, then we would have $R + T = 1$. According to Equation (4.13) this would give $\mathcal{T}_{max} = 1$. This means that the peak intensity of the transmitted fringes would be equal to the intensity of the incident light, even if R were very close to one. In actual practice A is never zero, and the maximum transmittance is always somewhat less than unity, namely

$$\mathcal{T}_{max} = \left(\frac{1 - A - R}{1 - R} \right)^2$$

(4.15)

4.2 The Fabry-Perot Interferometer

The interferometer devised by C. Fabry and A. Perot in 1899 employs multiple-beam interference. It is used to measure wavelengths with high precision and to study the fine structure of spectrum lines. A Fabry-Perot interferometer consists essentially of two optically flat,[1] partially reflecting plates of glass or quartz with their reflecting surfaces held accurately parallel. If the plate spacing can be mechanically varied, the device is called an *interferometer*, whereas if the plates are held fixed by spacers, it is called an *etalon*. The surfaces must be extremely flat and parallel in order to obtain the maximum fringe sharpness. An ordinary optical flat of $\frac{1}{4}$ wavelength flatness is not good enough for precise Fabry-Perot applications. Rather, a flatness of the order of 1/20 to 1/100 wavelength is required.

---

[1] The *spherical* Fabry-Perot interferometer, first introduced by Connes, employs spherical concave reflecting surfaces.

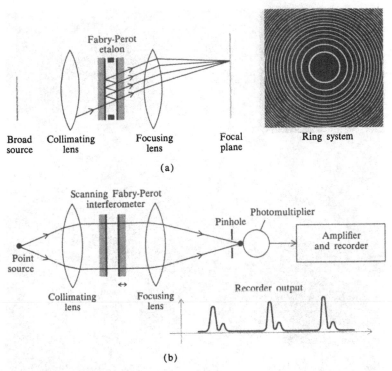

Figure 4.4. Arrangements for (a) the Fabry-Perot etalon; (b) the scanning interferometer.

In use, the interferometer is usually mounted between a collimating lens and a focusing lens as shown in Figure 4.4(a). If a broad source of light is used, interference fringes in the form of concentric circular rings appear in the focal plane of the focusing lens (Figure 4.5). These rings may be observed visually or photographed. A given ring corresponds to a constant value of θ, and the circular fringes are called *fringes of equal inclination*. Another way of using the interferometer, called the *scanning method*, employs a point source or a pinhole. The source is placed so that only one spot, the center of the ring system, appears at the exit focal plane as shown in Figure 4.4(b). Scanning may be accomplished by changing the spacing either mechanically or optically, say by changing the air pressure. The intensity at the ring center is usually recorded photoelectrically. This gives a graph of the interference pattern. The graphical record is essentially a plot of the Airy function $(1 + F \sin^2 \Delta/2)^{-1}$ or rather, a sum of such functions for each component frequency. A typical record is shown in Figure 4.4(b).

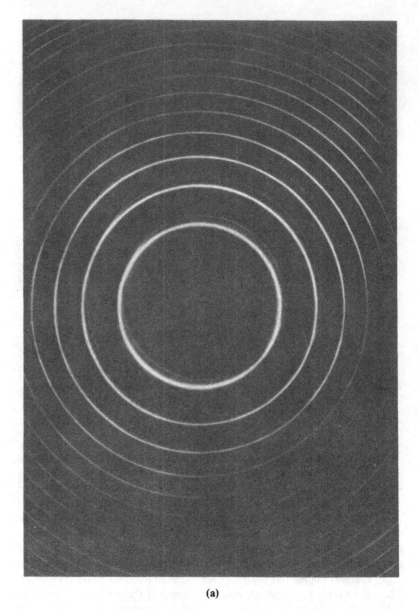

(a)

Figure 4.5.   Fabry-Perot interference fringes. (a) Monochromatic source; (b) non-monochromatic source.

(b)

The *free spectral range* of a Fabry-Perot instrument is defined as the separation between adjacent orders of interference. In terms of the parameter Δ, the free spectral range corresponds to

$$\Delta_{N+1} - \Delta_N = 2\pi$$

Thus from Equations (4.2) and (4.7) we have

$$\omega_{N+1} - \omega_N = \frac{\pi c}{nd \cos \theta}$$

or,

$$\nu_{N+1} - \nu_N = \frac{c}{2nd \cos \theta}$$

For small θ the free spectral range in frequency is approximately

$$\nu_{N+1} - \nu_N \cong \frac{c}{2nd} \tag{4.16}$$

4.3 Resolution of Fabry-Perot Instruments

Suppose a spectrum consisting of two closely spaced frequencies ω and ω' is to be analyzed with a Fabry-Perot interferometer. The intensity distribution will be a superposition of two fringe systems as indicated in Figure 4.6. Here the two components are assumed to be of equal intensity. The fringe pattern is given by the sum of two Airy functions:

$$I_T = I_0 \left(1 + F \sin^2 \frac{\Delta}{2} \right)^{-1} + I_0 \left(1 + F \sin^2 \frac{\Delta'}{2} \right)^{-1} \tag{4.17}$$

in which F is defined by Equation (4.9), and where

$$\Delta \approx \delta_r + 2kd = \delta_r + \frac{2\omega d}{c}$$

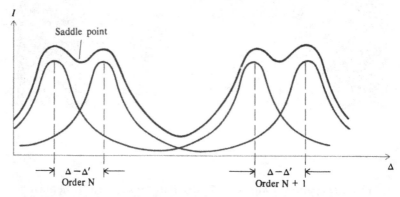

Figure 4.6. Graph of intensity distribution for two monochromatic lines in Fabry-Perot interferometry.

and similarly

$$\Delta' \approx \delta_r + 2k'd = \delta_r + \frac{2\omega'd}{c}$$

We assume that θ is small so that $\cos \theta \approx 1$.

Now the two frequencies ω and ω' can be said to be resolved if there is a dip in the intensity curve. A useful convention for resolution in the case of multiple-beam interference is known as the *Taylor criterion*. According to this convention, two equal lines are considered to be resolved if the individual curves cross at the half-intensity point, so that the total intensity at the saddle point is equal to the maximum intensity of either line alone. Thus at the saddle point, which is midway between the two frequencies, we can write

$$I = 2I_0 \left[1 + F \sin^2 \left(\frac{\Delta - \Delta'}{4} \right) \right]^{-1} = I_0$$

which gives

$$F \sin^2 \left(\frac{\Delta - \Delta'}{4} \right) = 1$$

If we now assume that the quantity $\Delta - \Delta'$ is small, so that the sine term can be replaced by its argument, we obtain

$$|\Delta - \Delta'| = 4F^{-1/2} = 2 \left(\frac{1-R}{\sqrt{R}} \right) \tag{4.18}$$

Finally, in terms of angular frequency we can write

$$\delta\omega = |\omega - \omega'| = \frac{2c}{d} F^{-1/2} = \frac{c}{d} \left(\frac{1-R}{\sqrt{R}} \right) \tag{4.19}$$

for the fringe width at half intensity or, equivalently, the resolution width of a Fabry-Perot interferometer for which the plate separation is d and the reflectance is R.

A parameter that is often used in interferometry is the ratio of the free spectral range to the fringe width, known as the *reflecting finesse* $\mathscr{F}$. It can be expressed in terms of the other parameters as follows:

$$\mathscr{F} = \frac{\Delta_{N+1} - \Delta_N}{|\Delta - \Delta'|} = \frac{\pi}{2} \sqrt{F} = \pi \left(\frac{\sqrt{R}}{1-R} \right) \tag{4.20}$$

The *resolving power*, abbreviated RP, of any spectroscopic instrument is defined as the reciprocal of the fractional resolution, namely,

$$RP = \frac{\omega}{\delta\omega} = \frac{\nu}{\delta\nu} = \frac{\lambda}{|\delta\lambda|}$$

in which the denominators refer to the minimum resolvable quantities (fringe widths) measured in angular frequency, frequency, or wavelength, respectively. From Equations (4.19) and (4.20) we see that the resolving power of a Fabry-Perot instrument can be expressed as

$$RP = N\mathscr{F} = N\pi\left(\frac{\sqrt{R}}{1 - R}\right) \tag{4.21}$$

and is thus determined by the order of interference and the reflectance.

By increasing the order of interference, the resolving power with a given reflectance can be made as large as desired. This is easily accomplished by increasing the mirror separation, because $N = 2nd/\lambda_0$, approximately, as given by Equation (4.10). However, the free spectral range then diminishes, and so a compromise must be chosen in any given application. For a given value of the mirror separation the resolving power, in principle, can be increased indefinitely by making the reflectance closer and closer to unity. However, a practical limit is imposed by absorption in the reflecting surface which reduces the intensity of the transmitted fringes, as discussed in Section 4.1.

Silver and aluminum films, deposited by vacuum evaporation, have long been used for Fabry-Perot instruments. The useful reflectance with these metal films, as limited by absorption, is only about 80 to 90 percent. More recently, multilayer dielectric films have been used for Fabry-Perot work. With such films, discussed in the next section, useful reflectances approaching 99 percent can be achieved. A good Fabry-Perot instrument can easily have a resolving power of 1 million, which is 10 to 100 times that of a prism or small-grating spectroscope. For a more complete discussion of Fabry-Perot interferometry, and of other instruments used in high-resolution spectroscopy, see References [5] and [41] listed at the end of the book.

4.4 Theory of Multilayer Films

Multilayer films are widely used in science and industry for control of light. Optical surfaces having virtually any desired reflectance and transmittance characteristics may be produced by means of thin film coatings. These films are usually deposited on glass or metal substrates by high-vacuum evaporation. The well-known use of anti-reflecting coatings for camera lenses and other optical instruments is only one of the many practical applications of thin films. Other applications include such things as heat-reflecting and heat-transmitting mirrors ("hot" and "cold" mirrors), one-way mirrors, optical filters, and so forth.

First consider the case of a single layer of dielectric of index n_1 and thickness l between two infinite media of indices n_0 and n_T (Figure 4.7). For simplicity we shall develop the theory for normally incident light. The modifications for the general case of oblique in-

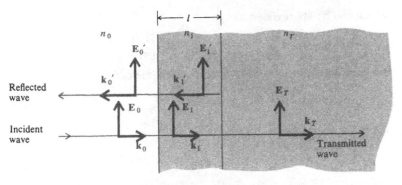

Figure 4.7. Wave vectors and their associated electric fields for the case of normal incidence on a single dielectric layer.

cidence are easily made. The amplitude of the electric vector of the incident beam is E_0. That of the reflected beam is E_0', and that of the transmitted beam is E_T. The electric-field amplitudes in the film are E_1 and E_1' for the forward and backward traveling waves, respectively, as indicated in the figure.

The boundary conditions require that the electric and magnetic fields be continuous at each interface. These conditions are expressed as follows:

First Interface	Second Interface
Electric: $E_0 + E_0' = E_1 + E_1'$	$E_1 e^{ikl} + E_1' e^{-ikl} = E_T$
Magnetic: $H_0 - H_0' = H_1 - H_1'$	$H_1 e^{ikl} - H_1' e^{-ikl} = H_T$
or $n_0 E_0 - n_0 E_0' = n_1 E_1 - n_1 E_1'$	$n_1 E_1 e^{ikl} - n_1 E_1' e^{-ikl} = n_T E_T$

The relations for the magnetic fields follow from the theory developed in Section 2.7. The phase factors e^{ikl} and e^{-ikl} result from the fact that the wave travels through a distance l from one interface to the other.

If we eliminate the amplitudes E_1 and E_1', we obtain

$$1 + \frac{E_0'}{E_0} = \left(\cos kl - i \frac{n_T}{n_1} \sin kl\right)\frac{E_T}{E_0}$$

$$n_0 - n_0 \frac{E_0'}{E_0} = (-i n_1 \sin kl + n_T \cos kl)\frac{E_T}{E_0}$$

(4.22)

or, in matrix form,

$$\begin{bmatrix} 1 \\ n_0 \end{bmatrix} + \begin{bmatrix} 1 \\ -n_0 \end{bmatrix}\frac{E_0'}{E_0} = \begin{bmatrix} \cos kl & \dfrac{-i}{n_1}\sin kl \\ -i n_1 \sin kl & \cos kl \end{bmatrix}\begin{bmatrix} 1 \\ n_T \end{bmatrix}\frac{E_T}{E_0}$$

(4.23)

which can be abbreviated as

$$\begin{bmatrix} 1 \\ n_0 \end{bmatrix} + \begin{bmatrix} 1 \\ -n_0 \end{bmatrix} r = M \begin{bmatrix} 1 \\ n_T \end{bmatrix} t \tag{4.24}$$

We have here introduced the reflection coefficient

$$r = \frac{E_0{'}}{E_0} \tag{4.25}$$

and the transmission coefficient

$$t = \frac{E_T}{E_0} \tag{4.26}$$

The matrix, known as the *transfer matrix*

$$M = \begin{bmatrix} \cos kl & -\dfrac{i}{n_1} \sin kl \\ -in_1 \sin kl & \cos kl \end{bmatrix} \tag{4.27}$$

where n_1 is the index of refraction, and $k = 2\pi/\lambda = 2\pi n_1/\lambda_0$.

Now suppose that we have N layers numbered 1, 2, 3, . . . N having indices of refraction n_1, n_2, n_3, . . . n_N and thicknesses l_1, l_2, l_3, . . . l_N, respectively. In the same way that we derived Equation (4.24), we can show that the reflection and transmission coefficients of the multilayer film are related by a similar matrix equation:

$$\begin{bmatrix} 1 \\ n_0 \end{bmatrix} + \begin{bmatrix} 1 \\ -n_0 \end{bmatrix} r = M_1 M_2 M_3 \cdots M_N \begin{bmatrix} 1 \\ n_T \end{bmatrix} t = M \begin{bmatrix} 1 \\ n_T \end{bmatrix} t \tag{4.28}$$

where the transfer matrices of the various layers are denoted by M_1, M_2, M_3, . . . , M_N. Each transfer matrix is of the form given by Equation (4.27) with appropriate values of n, l, and k. The overall transfer matrix M is the product of the individual transfer matrices. Let the elements of M be A, B, C, and D, that is

$$M_1 M_2 M_3 \cdots M_N = M = \begin{bmatrix} A & B \\ C & D \end{bmatrix} \tag{4.29}$$

We can then solve Equation (4.28) for r and t in terms of these elements. The result is

$$r = \frac{An_0 + Bn_T n_0 - C - Dn_T}{An_0 + Bn_T n_0 + C + Dn_T} \tag{4.30}$$

$$t = \frac{2n_0}{An_0 + Bn_T n_0 + C + Dn_T} \tag{4.31}$$

The reflectance R and the transmittance T are then given by $R = |r|^2$ and $T = |t|^2$, respectively.

Antireflecting Films The transfer matrix for a single film of index n_1 and thickness l is given by Equation (4.27). Suppose this film is placed on a glass substrate of index n_T. The coefficient of reflection of the combination, in air, is then given by Equation (4.30) with $n_0 = 1$. The result is

$$r = \frac{n_1(1 - n_T)\cos kl - i(n_T - n_1{}^2)\sin kl}{n_1(1 + n_T)\cos kl - i(n_T + n_1{}^2)\sin kl} \tag{4.32}$$

If the optical thickness of the film is $\frac{1}{4}$ wavelength, then $kl = \pi/2$. The reflectance for a quarter-wave film is therefore

$$R = |r|^2 = \frac{(n_T - n_1{}^2)^2}{(n_T + n_1{}^2)^2} \tag{4.33}$$

In particular, the reflectance is zero if

$$n_1 = \sqrt{n_T} \tag{4.34}$$

Magnesium fluoride, whose index is 1.35, is commonly used for coating lenses. Although this does not exactly satisfy the above requirement for ordinary glass, $n_T \approx 1.5$, the reflectance of glass coated with a quarter-wave layer of magnesium fluoride is reduced to about 1 percent, which is one fourth that of uncoated glass. This can result in a considerable saving of light in the case of optical instruments having many elements, such as high-quality camera lenses that may have as many as five or six components, that is, ten or twelve reflecting surfaces.

By using two layers, one of high index and one of low, it is possible to obtain zero reflectance (at one wavelength) with available coating materials. More layers obviously afford greater latitude and more extensive possibilities. Thus with three suitably chosen layers the reflectance can be reduced to zero for two wavelengths and can be made to average less than $\frac{1}{4}$ percent over almost the entire visible spectrum. Some curves are shown in Figure 4.8.

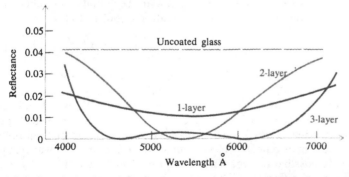

Figure 4.8. Curves of reflectance versus wavelength of antireflecting films.

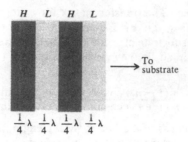

Figure 4.9. Multilayer stack for producing high reflectance. The stack consists of alternate quarter-wave layers of high and low index material. (Note: λ is the wavelength in the material.)

High-Reflectance Films In order to obtain a high value of reflectance in a multilayer film, a stack of alternate layers of high index, n_H, and low index, n_L, materials is used, the thickness of each layer being $\frac{1}{4}$ wavelength, Figure 4.9. In this case the transfer matrices are all of the same form, and the product of two adjacent ones is

$$\begin{bmatrix} 0 & \dfrac{-i}{n_L} \\ -in_L & 0 \end{bmatrix} \begin{bmatrix} 0 & \dfrac{-i}{n_H} \\ -in_H & 0 \end{bmatrix} = \begin{bmatrix} \dfrac{-n_H}{n_L} & 0 \\ 0 & \dfrac{-n_L}{n_H} \end{bmatrix} \tag{4.35}$$

If the stack consists of $2N$ layers, then the transfer matrix of the complete multilayer film is

$$M = \begin{bmatrix} \dfrac{-n_H}{n_L} & 0 \\ 0 & \dfrac{-n_L}{n_H} \end{bmatrix}^N = \begin{bmatrix} \left(\dfrac{-n_H}{n_L}\right)^N & 0 \\ 0 & \left(\dfrac{-n_L}{n_H}\right)^N \end{bmatrix} \tag{4.36}$$

Assuming, for simplicity, that n_0 and n are both unity, the reflectance of a multilayer stack is given by Equation (4.30) as follows:

$$R = |r|^2 = \left[\frac{(-n_H/n_L)^N - (-n_L/n_H)^N}{(-n_H/n_L)^N + (-n_L/n_H)^N}\right]^2 = \left[\frac{(n_H/n_L)^{2N} - 1}{(n_H/n_L)^{2N} + 1}\right]^2 \tag{4.37}$$

The reflectance thus approaches unity for large N. For instance, an eight-layer stack ($N = 4$) of zinc sulfide ($n_H = 2.3$) and magnesium fluoride ($n_L = 1.35$) gives a reflectance of about 0.97, which is higher than the reflectance of pure silver in the visible region of the spectrum. A 30-layer stack results in a reflectance of better than 0.999. This maximum reflectance, of course, occurs only at 1 wavelength, but it is possible to broaden the region of high reflectance by combinations of different thicknesses. Figure 4.10 shows some approx-

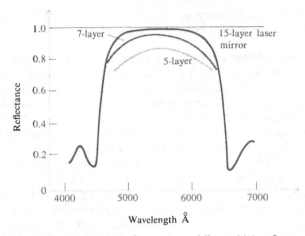

Figure 4.10. Reflectance curves for some multilayer high-reflectance films.

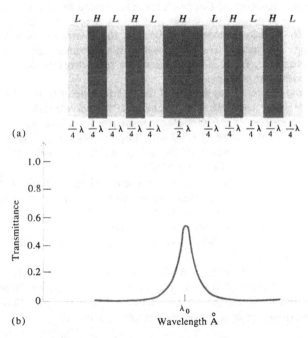

Figure 4.11. Multilayer Fabry-Perot interference filter.

102 MULTIPLE-BEAM INTERFERENCE

imate curves of reflectance as a function of wavelength for multilayer films such as are used in laser work.

Fabry-Perot Interference Filter A Fabry-Perot type of filter consists of a layer of dielectric having a thickness of $\frac{1}{2}$ wavelength (for some wavelength λ_0) and bounded on both sides by partially reflecting surfaces. In effect, this is a Fabry-Perot etalon with a very small spacing. The result is a filter that has a transmission curve given by the Airy function Equation (4.8), where the peak occurs at the wavelength λ_0. Higher-order peaks also occur for wavelengths $\frac{1}{2}\lambda_0$, $\frac{1}{3}\lambda_0$, $\frac{1}{4}\lambda_0$, . . . , and so forth. The spectral width of the transmission band depends on the reflectance of the bounding surfaces. Fabry-Perot filters were first made with silver films to produce the necessary high reflectance, but now they are usually made entirely of multilayer dielectric films. The latter are superior to the metal films because of their higher reflectance and lower absorption. Figure 4.11 shows a typical design of a multilayer Fabry-Perot interference filter together with a transmission curve.

PROBLEMS

4.1 The plates of a Fabry-Perot interferometer are coated with silver of such a thickness that for each plate the reflectance is 0.9, the transmittance is 0.05, and the absorption is 0.05. Find the maximum and minimum transmittance of the interferometer. What is the value of the reflecting finesse and of the coefficient of finesse?

4.2 Find the resolving power of the interferometer in Problem 4.1 if the plate separation is 1 cm and the wavelength is 500 nm.

4.3 The mirrors of a Fabry-Perot resonator for a laser are coated to give a reflectance of 0.99 and they are separated a distance of 1 m. Find the value of the fringe width in wavelength and in frequency at a wavelength of 633 nm.

4.4 Show that the radii of the interference fringes of a plane-parallel Fabry-Perot interferometer, as shown in Figure 4.5(a), are approximately proportional to $\sqrt{0}$, $\sqrt{1}$, $\sqrt{2}$, . . . $\sqrt{N'}$, where N' is an integer, provided that there is a zero-radius fringe. [Hint: Use Equation (4.10) and assume that θ is small, so that $\cos \theta \approx 1 - \theta^2/2$.]

4.5 A collimated beam of white light falls at normal incidence on a plate of glass of index n and thickness d. Develop a formula for the transmittance as a function of wavelength. Show that maxima occur at those wavelengths such that

$$\lambda_N = \frac{2nd}{N}$$

where λ_N is the vacuum wavelength and N is an integer. The transmission function is thus periodic (in wavenumber or frequency) and is called a "channeled spectrum."

4.6 Using the result of Problem 4.5, find the maximum and minimum transmittance of the channels for a plate of index of refraction 2.5. If the plate is 1 mm thick, find the wavelength separation between adjacent channels at a vacuum wavelength of 600 nm.

4.7 Calculate the reflectance of a quarter-wave antireflecting film of magnesium fluoride ($n = 1.35$) coated on an optical glass surface of index 1.52.

4.8 Find the peak reflectance of a high-reflecting multilayer film consisting of eight layers of high-low index material (four of each) of index $n_L = 1.4$ and $n_H = 2.8$.

4.9 Fill in the steps in the derivation of the expression of the transfer matrix of a single film

$$M = \begin{bmatrix} \cos kl & -\dfrac{i}{n_1} \sin kl \\ -in_1 \sin kl & \cos kl \end{bmatrix}$$

4.10 Show that in the case of oblique incidence the transfer matrix of a single film is

$$M = \begin{bmatrix} \cos \beta & -\dfrac{i}{p} \sin \beta \\ -ip \sin \beta & \cos \beta \end{bmatrix}$$

where

$$\beta = kl \cos \theta$$

and

$$p = n_1 \cos \theta \quad (TE \text{ polarization})$$

$$p = \frac{n_1}{\cos \theta} \quad (TM \text{ polarization})$$

The angle θ is the angle between the wave vector inside and the normal to the surface of the film.

CHAPTER 5
Diffraction

5.1 General Description of Diffraction

If an opaque object is placed between a point source of light and a white screen, it is found that the shadow that is cast by the object departs from the perfect sharpness predicted by geometrical optics. Close examination of the shadow edge reveals that some light goes over into the dark zone of the geometrical shadow and that dark fringes appear in the illuminated zone. This "smearing" of the shadow edge is closely related to another phenomenon, namely, the spreading of light after passing through a very small aperture, such as a pinhole or a narrow slit, as in Young's experiment. The collective name given to these departures from geometrical optics is *diffraction.*

The essential features of diffraction phenomena can be explained qualitatively by Huygens' principle. This principle in its original form states that the propagation of a light wave can be predicted by assuming that each point of the wave front acts as the source of a secondary wave that spreads out in all directions. The envelope of all the secondary waves is the new wave front.

We shall not attempt to treat diffraction by a direct application of Huygens' principle. We want a more quantitative approach. Our strategy will be to cast Huygens' principle into a precise mathematical form known as the *Fresnel-Kirchhoff formula.* This formula will then be applied to various specific cases of diffraction of light by obstacles and apertures.

5.2 Fundamental Theory

Let us recall Green's theorem[1] that states that if U and V are any two scalar-point functions that satisfy the usual conditions of continuity and integrability, then the following equality holds:

$$\iint (V \operatorname{grad}_n U - U \operatorname{grad}_n V)\, d\mathscr{A} = \iiint (V\, \nabla^2 U - U\, \nabla^2 V)\, d\mathscr{V} \qquad (5.i)$$

---

[1] Green's theorem can be proved from the divergence theorem $\iint \operatorname{grad}_n \mathbf{F}\, d\mathscr{A} = \iiint \nabla \cdot \mathbf{F}\, d\mathscr{V}$ by setting $\mathbf{F} = U\, \nabla V - V\, \nabla U$ and using the vector identity $\nabla \cdot (U\, \nabla V) = U\, \nabla^2 V + (\nabla U) \cdot (\nabla V)$.

106

The left-hand integral extends over any closed surface $\mathscr{A}$, and the right-hand integral includes the volume $\mathscr{V}$ within that surface. By "grad_n" is meant the normal component of the gradient at the surface of integration.

In particular, if both U and V are wave functions; that is, if they satisfy the regular wave equations

$$\nabla^2 U = \frac{1}{u^2}\frac{\partial^2 U}{\partial t^2} \qquad \nabla^2 V = \frac{1}{u^2}\frac{\partial^2 V}{\partial t^2}$$

and if they both have a harmonic time dependence of the form $e^{\pm i\omega t}$, then it is straightforward to show that the volume integral in Green's theorem is identically zero. The theorem then reduces to

$$\iint (V \text{ grad}_n U - U \text{ grad}_n V)\, d\mathscr{A} = 0 \qquad (5.2)$$

Now suppose that we take V to be the wave function

$$V = V_0 \frac{e^{i(kr+\omega t)}}{r} \qquad (5.3)$$

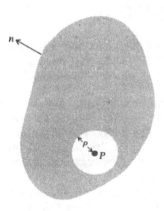

Figure 5.1. Surface of integration for proving the Kirchhoff integral theorem.

This particular function represents spherical waves converging to the point P ($r = 0$). We let the volume enclosed by the surface of integration include the point P. Since V becomes infinite at P, we must exclude that point from the integration. This is accomplished by the standard method of subtracting an integral over a small sphere of radius ρ centered at P, as indicated in Figure 5.1. Over this small sphere, $r = \rho$ and $\text{grad}_n = -\partial/\partial r$. Hence we can write

$$\iint \left(\frac{e^{ikr}}{r} \, \text{grad}_n \, U - U \, \text{grad}_n \, \frac{e^{ikr}}{r} \right) d\mathscr{A}$$

$$- \iint \left(\frac{e^{ikr}}{r} \frac{\partial U}{\partial r} - U \frac{\partial}{\partial r} \frac{e^{ikr}}{r} \right)_{r=\rho} \rho^2 \, d\Omega = 0 \qquad (5.4)$$

where $d\Omega$ is the element of solid angle on the sphere centered at P, and $\rho^2 \, d\Omega$ is the corresponding element of area. The common factor $V_0 e^{i\omega t}$ has been canceled out.

We now let ρ shrink to zero. Then, in the limit as ρ approaches zero the integrand of the second integral approaches the value that U has at the point P, namely U_p. This is easily verified by performing the indicated operations. Consequently, the second integral itself, including the sign, approaches the value

$$\iint U_p \, d\Omega = 4\pi U_p \qquad (5.5)$$

Equation (5.4) then becomes, on rearranging terms,

$$U_p = -\frac{1}{4\pi} \iint \left(U \, \text{grad}_n \, \frac{e^{ikr}}{r} - \frac{e^{ikr}}{r} \, \text{grad}_n \, U \right) d\mathscr{A} \qquad (5.6)$$

This equation is known as the *Kirchhoff integral theorem*. It relates the value of any scalar wave function at any point P *inside* an arbitrary closed surface to the value of the wave function *at* the surface.

In the application of Kirchhoff's theorem to diffraction, the wave function U is known as the "optical disturbance." Being a scalar quantity, it cannot accurately represent an electromagnetic field. However, in this so-called "scalar approximation" the square of the absolute value of U may be regarded as a measure of the irradiance at a given point.

The more rigorous theory of diffraction, which takes into account the vectorial nature of light, is beyond the scope of this book. Owing to the mathematical complexity of the rigorous theory, complete calculations have been carried out for only a relatively few simple cases [5].

The Fresnel-Kirchhoff Formula We now proceed to apply the Kirchhoff integral theorem to the general problem of diffraction of light. The diffraction is produced by an aperture of arbitrary shape in an otherwise opaque partition. This partition separates a light source from a receiving point (Figure 5.2).

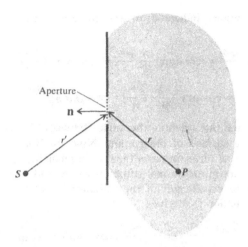

Figure 5.2. Geometry for the Fresnel-Kirchhoff formula.

Our task is to determine the optical disturbance reaching the receiving point P from the source S. In applying the Kirchhoff integral we choose a surface of integration such that it encloses the receiving point and includes, as a part of it, the aperture opening as indicated in the figure.

Two basic simplifying assumptions are introduced:

(1) The wave function U and its gradient contribute negligible amounts to the integral except at the aperture opening itself.
(2) The values of U and grad U at the aperture are the same as they would be in the absence of the partition.

Although the validity of these assumptions is open to considerable debate, the results are generally in good agreement with experimental observations.

If r' denotes the position of a point on the aperture relative to the source S, then the wave function at the aperture is given by the expression

$$U = U_o \frac{e^{i(kr'-\omega t)}}{r'} \qquad (5.7)$$

which represents spherical monochromatic waves traveling outward from S. The Kirchhoff integral theorem then yields

$$U_\mathrm{p} = \frac{U_0 e^{-i\omega t}}{4\pi} \iint \left(\frac{e^{ikr}}{r} \,\mathrm{grad}_\mathrm{n}\, \frac{e^{ikr'}}{r'} - \frac{e^{ikr'}}{r'} \,\mathrm{grad}_\mathrm{n}\, \frac{e^{ikr}}{r} \right) d\mathscr{A} \qquad (5.8)$$

where the integration extends only over the aperture opening.

The operations indicated in the integrand are carried out as follows:

$$\text{grad}_n\ \frac{e^{ikr}}{r} = \cos\ (\mathbf{n,r})\ \frac{\partial}{\partial r}\ \frac{e^{ikr}}{r} = \cos\ (\mathbf{n,r})\left(\frac{ike^{ikr}}{r} - \frac{e^{ikr}}{r^2}\right) \quad (5.9)$$

$$\text{grad}_n\ \frac{e^{ikr'}}{r'} = \cos\ (\mathbf{n,r'})\ \frac{\partial}{\partial r'}\ \frac{e^{ikr'}}{r'} = \cos\ (\mathbf{n,r'})\left(\frac{ike^{ikr'}}{r'} - \frac{e^{ikr''}}{r'^2}\right) \quad (5.10)$$

where $(\mathbf{n,r'})$ and $(\mathbf{n,r'})$ denote the angles between the vectors and the normal to the surface of integration. Now, in Equations (5.9) and (5.10) the second terms in parentheses are negligibly small compared to the first terms in the normal situation where both r and r' are much larger than the wavelength of the radiation because $k = 2\pi/\lambda$. Consequently Equation (5.8) gives

$$U_p = -\frac{ikU_0e^{-i\omega t}}{4\pi}\iint \frac{e^{ik(r+r')}}{rr'}\ [\cos\ (\mathbf{n,r}) - \cos\ (\mathbf{n,r'})]\ d\mathscr{A} \quad (5.11)$$

This equation is known as the *Fresnel Kirchhoff integral formula*. It is, in effect, a mathematical statement of Huygens' principle. This is most easily seen by applying the formula to a specific case, namely, that of a circular aperture with the source symmetrically located as shown in Figure 5.3. The surface of integration is taken to be a spher-

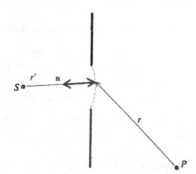

Figure 5.3. Diagram to show how Huygens' principle follows from the Kirchhoff integral formula.

ical cap bounded by the aperture opening. In this case r' is constant and $\cos\ (\mathbf{n,r'}) = -1$. The Fresnel-Kirchhoff formula then reduces to

$$U_p = -\frac{ik}{4\pi}\iint \frac{U_\mathscr{A}e^{i(kr-\omega t)}}{r}\ [\cos\ (\mathbf{n,r}) + 1]\ d\mathscr{A} \quad (5.12)$$

where

$$U_\mathscr{A} = \frac{U_0e^{ikr'}}{r'}$$

Equation (5.12) can be given the following simple interpretation: $U_\mathscr{A}$ is the complex amplitude of the incident primary wave at the aperture. From this primary wave each element $d\mathscr{A}$ of the aperture gives rise to a secondary spherical wave

$$\frac{U_\mathscr{A} e^{i(kr-\omega t)}}{r} d\mathscr{A}$$

The total optical disturbance at the receiving point P is obtained by summing the secondary waves from each element. However, in the summation it is necessary to take into account the factor cos $(\mathbf{n},\mathbf{r})$ − cos $(\mathbf{n},\mathbf{r}')$ known as the *obliquity factor*. In the case under discussion cos $(\mathbf{n},\mathbf{r}') = -1$, so the obliquity factor is cos $(\mathbf{n},\mathbf{r}) + 1$. In the forward direction cos $(\mathbf{n},\mathbf{r}) = 1$, and the obliquity factor is then equal to 2, its maximum value. On the other hand, in the backward direction cos $(\mathbf{n},\mathbf{r}) = -1$, so the obliquity factor is zero. This explains why there is no backward progressing wave created by the original wave front. Huygens' principle, as originally proposed, did not include the obliquity factor and thus could not account for the absence of a backward wave. The presence of the factor $-i$ means that the diffracted waves are shifted in phase by 90 degrees with respect to the primary incident wave. This feature was also lacking in the original form of Huygens' principle.

Complementary Apertures. Babinet's Principle Consider a diffracting aperture $\mathscr{A}$ that produces a certain optical disturbance U_p at a given observing point P. Suppose, now, that the aperture is divided into two portions $\mathscr{A}_1$ and $\mathscr{A}_2$ such that $\mathscr{A} = \mathscr{A}_1 + \mathscr{A}_2$. The two apertures $\mathscr{A}_1$ and $\mathscr{A}_2$ are then said to be *complementary*. An example is shown in Figure 5.4. From the form of the Fresnel-Kirchhoff formula, it is clear that

$$U_p = U_{1p} + U_{2p} \tag{5.13}$$

where U_{1p} is the optical disturbance at P produced by aperture $\mathscr{A}_1$ alone, and U_{2p} is the optical disturbance produced by aperture $\mathscr{A}_2$ alone. The equation above is one form of a theorem known as *Babinet's principle*.

Babinet's principle is useful in certain special cases. For instance, if $U_p = 0$, then $U_{1p} = -U_{2p}$. The complementary apertures in this case yield identical optical disturbances, except that they differ in phase by 180 degrees. The intensity at P, being equal to the absolute square of the optical disturbance, is therefore the same for the two apertures. Thus a collimated beam of light, such as that from a searchlight, will undergo diffraction scattering from a small spherical particle or a large number of particles such as fog. The condition $U_p = 0$ is then valid for those points P not in the direct beam, the

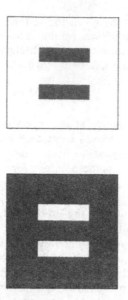

Figure 5.4. Complementary apertures.

aperture being defined by the size of the beam itself. Then, according to Babinet's principle, the same diffraction pattern will result if the beam is diffused by a screen containing a small circular hole or a large number of randomly placed holes the same size as the fog particles.

5.3   Fraunhofer and Fresnel Diffraction

In the detailed treatment of diffraction it is customary to distinguish between two general cases. These are known as *Fraunhofer diffraction* and *Fresnel diffraction*. Qualitatively speaking, Fraunhofer dif-

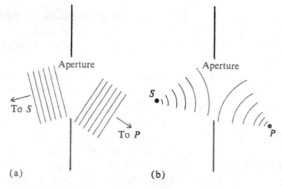

Figure 5.5. Diffraction by an aperture. (a) Fraunhofer case; (b) Fresnel case.

fraction occurs when both the incident and diffracted waves are effectively plane. This will be the case when the distances from the source
to the diffracting aperture and from the aperture to the receiving point
are both large enough for the curvatures of the incident and diffracted
waves to be neglected [Figure 5.5 (a)].

If either the source or the receiving point is close enough to the
diffracting aperture so that the curvature of the wave front is significant, then one has Fresnel diffraction [Figure 5.5(b)]. There is, of
course, no sharp line of distinction between the two cases. However,
a quantitative criterion can be obtained as follows. Consider Figure
5.6, which shows the general geometry of the diffraction problem.

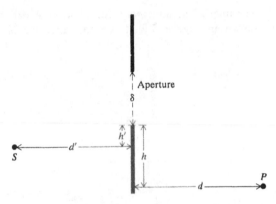

Figure 5.6. Geometry to show distinction between Fraunhofer diffraction
and Fresnel diffraction.

The receiving point P is located a distance d from the plane of the
diffracting aperture, and the source S is a distance d' from this plane.
One edge of the aperture is located a distance h from the foot of the
perpendicular drawn from P to the plane of the aperture. The corresponding distance for the source is h' as shown. The size of the aperture opening is δ. From the figure it is seen that the variation Δ of the
quantity $r + r'$ from one edge of the aperture to the other is given by

$$\Delta = \sqrt{d'^2 + (h' + \delta)^2} + \sqrt{d^2 + (h + \delta)^2} - \sqrt{d'^2 + h'^2} - \sqrt{d^2 + h^2}$$

$$= \left(\frac{h'}{d'} + \frac{h}{d}\right)\delta + \frac{1}{2}\left(\frac{1}{d'} + \frac{1}{d}\right)\delta^2 + \cdots \qquad (5.14)$$

The quadratic term in the expansion above is essentially a measure of
the curvature of the wave front. The wave is effectively plane over
the aperture if this term is negligibly small compared to the wavelength of the light, that is, if

$$\frac{1}{2}\left(\frac{1}{d'} + \frac{1}{d}\right)\delta^2 \ll \lambda \qquad (5.15)$$

This is the criterion for Fraunhofer diffraction. If this condition does not obtain, the curvature of the wave front becomes important and the diffraction is of the Fresnel type. Similar considerations apply in the case of diffraction by an opaque object or obstacle. Then δ is the linear size of the object. (Note that Babinet's principle applies here.)

Examples of Fraunhofer and Fresnel diffraction by various types of apertures are treated in the sections that follow. Since the Fraunhofer case is, in general, mathematically simpler than the Fresnel case, Fraunhofer diffraction will be discussed first.

5.4   Fraunhofer Diffraction Patterns

The usual experimental arrangement for observing Fraunhofer diffraction is shown in Figure 5.7. Here the aperture is *coherently*

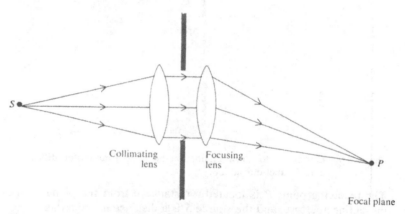

Figure 5.7.   Arrangement for observing Fraunhofer diffraction.

illuminated by means of a point monochromatic source and a collimating lens. A second lens is placed behind the aperture as shown. The incident and diffracted wave fronts are therefore strictly plane, and the Fraunhofer case is rigorously valid. In applying the Fresnel-Kirchhoff formula [Equation (5.11)] to the calculation of the diffraction patterns, the following simplifying approximations are taken to be valid:

(1) The angular spread of the diffracted light is small enough for the obliquity factor [cos (**n**,**r**) − cos (**n**,**r**′)] not to vary appreciably over the aperture and to be taken outside the integral.

(2) The quantity $e^{ikr'}/r'$ is very nearly constant and can be taken outside the integral.

(3) The variation of the remaining factor e^{ikr}/r over the aperture comes principally from the exponential part, so the factor $1/r$ can be replaced by its mean value and taken outside the integral.

Consequently, the Fresnel-Kirchhoff formula reduces to the very simple equation

$$U_p = C \iint e^{ikr}\, d\mathscr{A} \tag{5.16}$$

where all constant factors have been lumped into one constant C. The formula above states that the distribution of the diffracted light is obtained simply by integrating the phase factor e^{ikr} over the aperture.

The Single Slit   The case of diffraction by a single narrow slit is treated here as a one-dimensional problem. Let the slit be of length L

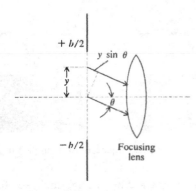

Figure 5.8.   Definition of the variables for Fraunhofer diffraction by a single slit.

and of width b. The element of area is then $d\mathscr{A} = L\, dy$ as indicated in Figure 5.8. Furthermore, we can express r as

$$r = r_0 + y \sin \theta \tag{5.17}$$

where r_0 is the value of r for $y = 0$, and where θ is the angle shown. The diffraction formula (5.16) then yields

$$U = C e^{ikr_0} \int_{-b/2}^{+b/2} e^{iky \sin \theta} L\, dy$$

$$= 2\, C e^{ikr_0} L\, \frac{\sin\left(\tfrac{1}{2} kb \sin \theta\right)}{k \sin \theta} = C' \left(\frac{\sin \beta}{\beta}\right) \tag{5.18}$$

where $\beta = \tfrac{1}{2} kb \sin \theta$, and $C' = e^{ikr_0} CbL$ is merely another constant.

Thus C' (sin β/β) is the total amplitude of the light diffracted in a given direction defined by β. This light is brought to a focus by the second lens, and the corresponding irradiance distribution in the focal plane is given by the expression

$$I = |U|^2 = I_0 \left(\frac{\sin \beta}{\beta}\right)^2 \qquad (5.19)$$

where $I_0 = |CLb|^2$, which is the irradiance for $\theta = 0$. The distribution is plotted in Figure 5.9. The maximum value occurs at $\theta = 0$, and zero

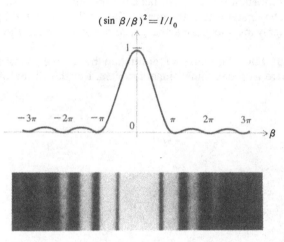

Figure 5.9. Fraunhofer diffraction pattern of a single slit.

values occur for $\beta = \pm\pi, \pm 2\pi, \ldots$, and so forth. Secondary maxima of rapidly diminishing value occur between these zero values. Thus the diffraction pattern at the focal plane consists of a central bright band. On either side there are alternating bright and dark bands. Table 5.1 gives the relative values of I of the first three sec-

Table 5.1. RELATIVE VALUES OF THE MAXIMA OF DIFFRACTION PATTERNS OF RECTANGULAR AND CIRCULAR APERTURES

	Rectangular	*Circular*
Central Max	1	1
1st Max	0.0496	0.0174
2d Max	0.0168	0.0042
3rd Max	0.0083	0.0016

ondary maxima. The first minimum, $\beta = \pi$, corresponds to

$$\sin \theta = \frac{2\pi}{kb} = \frac{\lambda}{b} \tag{5.20}$$

Thus, for a given wavelength, the angular width of the diffraction pattern varies inversely with the slit width, and the amplitude of the central maximum is proportional to the area of the slit. For very narrow slits the pattern is dim but wide. It shrinks and becomes brighter as the slit is widened.

The Rectangular Aperture The case of diffraction by a single aperture of rectangular shape is treated in the same way as the single slit, except that one must now integrate in two dimensions, say x and y as shown in Figure 5.10. It is left as a problem to show that the ir-

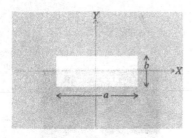

Figure 5.10.   Rectangular aperture.

radiance distribution is given by the product of two single-slit distribution functions. (See Section 5.6.) The result is

$$I = I_0 \left(\frac{\sin \alpha}{\alpha} \right)^2 \left(\frac{\sin \beta}{\beta} \right)^2 \tag{5.21}$$

where $\alpha - \frac{1}{2}ka \sin \phi$, $\beta = \frac{1}{2}kb \sin \theta$. The dimensions of the aperture are a and b and the angles ϕ and θ define the direction of the diffracted ray. The resulting diffraction pattern (Figure 5.11) has lines of zero irradiance defined by $\alpha = \pm\pi, \pm 2\pi, \ldots,$ and $\beta = \pm\pi, \pm 2\pi \ldots.$ As with the slit, the scale of the diffraction pattern bears an inverse relationship to the scale of the aperture.

The Circular Aperture To calculate the diffraction pattern of a circular aperture, we choose y as the variable of integration, as in the case of the single slit. If R is the radius of the aperture, then the element of area is taken to be a strip of width dy and length $2\sqrt{R^2 - y^2}$ (Figure 5.12).

The amplitude distribution of the diffraction pattern is then given by

$$U = C e^{ikr_0} \int_{-R}^{+R} e^{iky \sin \theta} \, 2\sqrt{R^2 - y^2} \, dy \tag{5.22}$$

$$\left(\frac{\sin\ \alpha}{\alpha}\right)^2\left(\frac{\sin\ \beta}{\beta}\right)^2 = I/I_0$$

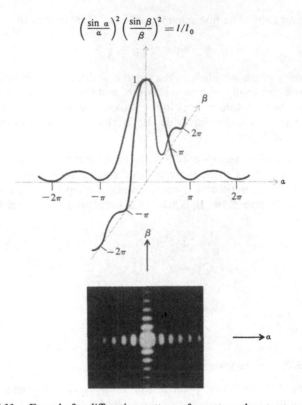

Figure 5.11. Fraunhofer diffraction pattern of a rectangular aperture.

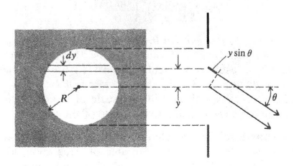

Figure 5.12. Circular aperture.

We introduce the quantities u and ρ defined by $u = y/R$ and $\rho = kR \sin\theta$. The integral in Equation 5.22 then becomes

$$\int_{-1}^{+1} e^{i\rho u}\ \sqrt{1 - u^2}\ du \qquad (5.23)$$

This is a standard integral. Its value is $\pi J_1(\rho)/\rho$ where J_1 is the Bessel function of the first kind, order one [27]. The ratio $J_1(\rho)/\rho \to \tfrac{1}{2}$ as $\rho \to 0$. The irradiance distribution is therefore given by

$$I = I_0 \left[\frac{2J_1(\rho)}{\rho} \right]^2 \tag{5.24}$$

where $I_0 = (C\pi R^2)^2$, which is the intensity for $\theta = 0$.

A graph of the intensity function is shown in Figure 5.13. The diffraction pattern is circularly symmetric and consists of a bright cen-

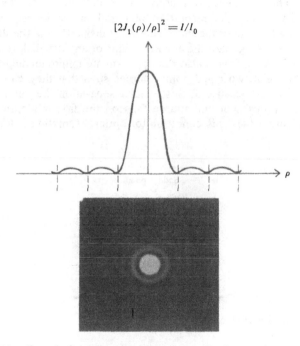

Figure 5.13.   Fraunhofer diffraction pattern of a circular aperture.

tral disk surrounded by concentric circular bands of rapidly diminishing intensity. The bright central area is known as the *Airy disk*. It extends to the first dark ring whose size is given by the first zero of the Bessel function, namely, $\rho = 3.832$. The angular radius of the first dark ring is thus given by

$$\sin \theta = \frac{3.832}{kR} = \frac{1.22\lambda}{D} \approx \theta \tag{5.25}$$

which is valid for small values of θ. Here $D = 2R$ is the diameter of the aperture.

The angular size of the Airy disk is thus slightly larger than the corresponding value λ/b for the bright central band of the diffraction pattern of the rectangular aperture or slit. In Table 5.1 are listed the values of the first few maxima of the diffraction patterns of rectangular and circular apertures.

Optical Resolution The image of a distant point source formed at the focal plane of an optical-telescope lens or a camera lens is actually a Fraunhofer diffraction pattern for which the aperture is the lens opening. Thus the image of a composite source is a superposition of many Airy disks. The resolution of detail in the image therefore depends on the size of the individual Airy disks. If D is the diameter of the lens opening, then the angular radius of an Airy disk is approximately $1.22 \lambda/D$. This is also the approximate minimum angular separation between two equal point sources such that they can be just barely resolved, because at this angular separation the central maximum of the image of one source falls on the first minimum of the other (Figure 5.14). This condition for optical resolution is known as

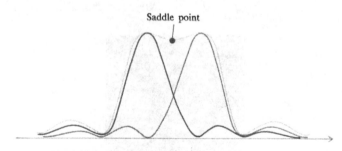

Saddle point

Figure 5.14. Rayleigh criterion.

the *Rayleigh criterion*. It is more convenient to use, in this case, than the Taylor criterion mentioned earlier.

In the case of the rectangular aperture, the minimum angular separation according to the Rayleigh criterion is just λ/b, where b is the width of the aperture. The intensity at the saddle point in this case is $8/\pi^2 = 0.81$ times the maximum intensity. The proof of this statement is left as an exercise.

The Double Slit Let us consider a diffracting aperture consisting of two parallel slits, each of width b and separated by a distance h (Figure 5.15). As with the single slit, we treat this case as a one-

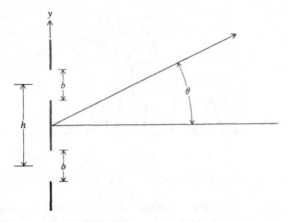

Figure 5.15. Double-slit aperture.

dimensional problem. The relevant diffraction integral is evaluated as follows:

$$\int_{\mathscr{A}} e^{iky\sin\theta}\,dy = \int_0^b e^{iky\sin\theta}\,dy + \int_h^{h+b} e^{iky\sin\theta}\,dy$$

$$= \frac{1}{ik\sin\theta}\left(e^{ikb\sin\theta} - 1 + e^{ik(h+b)\sin\theta} - e^{ikh\sin\theta}\right)$$

$$= \left(\frac{e^{ikb\sin\theta} - 1}{ik\sin\theta}\right)\left(1 + e^{ikh\sin\theta}\right) \qquad (5.26)$$

$$= 2b\,e^{i\beta}\,e^{i\gamma}\,\frac{\sin\beta}{\beta}\cos\gamma$$

where $\beta = \frac{1}{2}kb\sin\theta$ and $\gamma = \frac{1}{2}kh\sin\theta$. The corresponding irradiance distribution function is

$$I = I_0\left(\frac{\sin\beta}{\beta}\right)^2\cos^2\gamma \qquad (5.27)$$

The factor $(\sin\beta/\beta)^2$ is the previously found distribution function for a single slit. Here this factor constitutes an envelope for the interference fringes given by the term $\cos^2\gamma$. A plot is shown in Figure 5.16. Bright fringes occur for $\gamma = 0, \pm\pi, \pm2\pi$, and so forth. The angular separation between fringes is given by $\Delta\gamma = \pi$, or, approximately, in terms of the angle θ

$$\Delta\theta \approx \frac{2\pi}{kh} = \frac{\lambda}{h} \qquad (5.28)$$

It is interesting to note that this is equivalent to the result of the analysis of Young's experiment [Equation (3.9)].

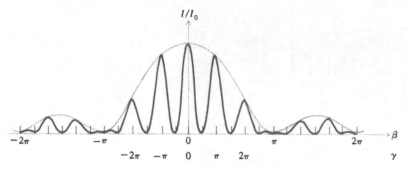

Figure 5.16.   Fraunhofer diffraction pattern of a double-slit aperture.

Multiple Slits. Diffraction Gratings   Let the aperture consist of a grating, that is, a large number N of identical parallel slits of width b and separation h (Figure 5.17). The evaluation of the diffractional integral is carried out in a manner similar to that of the double slit:

$$
\begin{aligned}
\int_{\mathcal{A}} e^{iky \sin \theta} \, dy &= \int_0^b + \int_h^{h+b} + \int_{2h}^{2h+b} + \cdots + \int_{(N-1)h}^{(N-1)h+b} e^{iky \sin \theta} \, dy \\
&= \frac{e^{ikb \sin \theta} - 1}{ik \sin \theta} \left[1 + e^{ikh \sin \theta} + \cdots + e^{ik(N-1)h \sin \theta} \right] \\
&= \frac{e^{ikb \sin \theta} - 1}{ik \sin \theta} \cdot \frac{1 - e^{ikNh \sin \theta}}{1 - e^{ikh \sin \theta}} \\
&= b e^{i\beta} e^{i(N-1)\gamma} \left(\frac{\sin \beta}{\beta} \right) \left(\frac{\sin N\gamma}{\sin \gamma} \right)
\end{aligned}
\tag{5.29}
$$

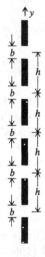

Figure 5.17.   Multiple-slit aperture or diffraction grating.

where $\beta = \frac{1}{2}kb \sin \theta$ and $\gamma = \frac{1}{2}kh \sin \theta$. This yields the following intensity distribution function:

$$I = I_0 \left(\frac{\sin \beta}{\beta}\right)^2 \left(\frac{\sin N\gamma}{N \sin \gamma}\right)^2 \qquad (5.30)$$

The factor N has been inserted in order to normalize the expression. This makes $I = I_0$ when $\theta = 0$.

Again the single-slit factor $(\sin \beta/\beta)^2$ appears as the envelope of the diffraction pattern. Principal maxima occur within the envelope at $\gamma = n\pi$, $n = 0, 1, 2, \ldots$, that is,

$$n\lambda = h \sin \theta \qquad (5.31)$$

which is the grating formula giving the relation between wavelength and angle of diffraction. The integer n is called the *order of diffraction*.

Secondary maxima occur near $\gamma = 3\pi/2N$, $5\pi/2N$, and so forth, and zeros occur at $\gamma = \pi/N$, $2\pi/N$, $3\pi/N$. . . . A graph is shown in Figure 5.18(a). If the slits are very narrow, then the factor $\sin \beta/\beta \approx 1$. The first few primary maxima, then, all have approximately the same value, namely, I_0.

Resolving Power of a Grating The angular width of a principal fringe, that is, the separation between the peak and the adjacent minimum, is found by setting the *change* of the quantity $N\gamma$ equal to π, that is, $\Delta\gamma = \pi/N = \frac{1}{2}kh \cos \theta \, \Delta\theta$, or

$$\Delta\theta = \frac{\gamma\lambda}{Nh \cos \theta} \qquad (5.32)$$

Thus if N is made very large, then $\Delta\theta$ is very small, and the diffraction pattern consists of a series of sharp fringes corresponding to the different orders $n = 0, \pm1, \pm2$, and so forth [Figure 5.18(b), (c)]. On the other hand for *a given order* the dependence of θ on the wavelength [Equation (5.31)] gives by differentiation

$$\Delta\theta = \frac{n \, \Delta\lambda}{h \cos \theta} \qquad (5.33)$$

This is the angular separation between two spectral lines differing in wavelength by $\Delta\lambda$. Combining Equation (5.32) and (5.33), we obtain the *resolving power* of a grating spectroscope according to the Rayleigh criterion, namely,

$$RP = \frac{\lambda}{\Delta\lambda} = Nn \qquad (5.34)$$

In words, the resolving power is equal to the number of grooves N multiplied by the order number n.

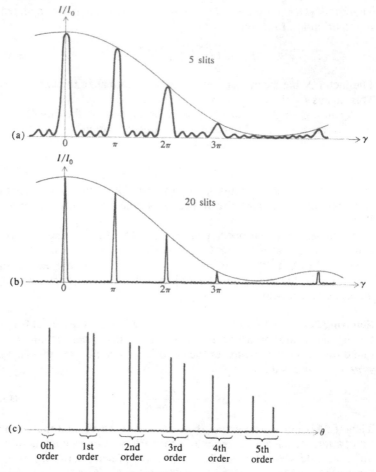

Figure 5.18. Fraunhofer diffraction pattern of a multiple-slit aperture. Graphs (a) and (b) are for monochromatic light. Graph (c) shows the pattern for a many-line grating illuminated with two different wavelengths.

Diffraction gratings used for optical spectroscopy are made by ruling grooves on a transparent surface (transmission type) or on a metal surface (reflection type). A typical grating may have, say, 600 lines/mm ruled over a total width of 10 cm. This would give a total of 60,000 lines and a theoretical resolving power of 60,000 n, where n is the order of diffraction used. In practice, resolving powers up to 90 percent of the theoretical values are obtainable with good gratings.

If the grooves are suitably shaped, usually of a sawtooth profile, most of the diffracted light can be made to appear in one order, thus increasing the efficiency of the grating. The essential requirement is that the spacing be uniform, within a fraction of a wavelength. This places extreme requirements on the mechanical rigidity of the ruling machine. High-quality replica gratings can be produced by a plastic molding process. These are much less expensive than original gratings.

Most of the gratings used in practical spectroscopy are of the reflection type. Reflection gratings are made with the ruled surface either plane or concave (Figure 5.19). Plane gratings require the use

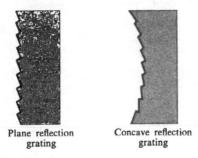

Plane reflection   Concave reflection
grating   grating

Figure 5.19. Reflection gratings.

of collimating and focusing lenses or mirrors, whereas concave gratings can perform the collimating and focusing functions as well as disperse the light into a spectrum. For more information on the subject of diffraction gratings and their use, the reader should consult References [17] and [35].

5.5 Fresnel Diffraction Patterns

According to the criteria discussed in Section 5.3, diffraction is of the Fresnel type when either the light source or the observing screen, or both, are so close to the diffracting aperture that the curvature of the wave front becomes significant. Since one is no longer dealing with plane waves, Fresnel diffraction is mathematically more difficult to treat than Fraunhofer diffraction but is actually simpler to observe experimentally because all that is needed is a source of light, an observing screen, and the diffracting aperture. The previously mentioned fringe effects seen around shadows are examples of Fresnel diffraction. In this section we shall discuss only a few relatively simple cases of Fresnel diffraction, which can be handled by elementary mathematical methods.

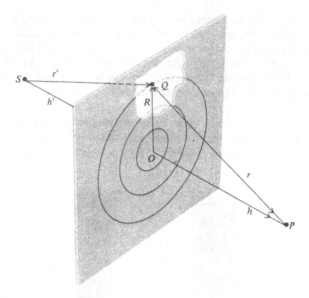

Figure 5.20. Fresnel zones in a plane aperture.

Fresnel Zones Consider a plane aperture illuminated by a point source S (Figure 5.20) such that a straight line connecting S to the receiving point P is perpendicular to the plane of the aperture. Let O be the point of intersection of the line SP with the aperture plane, and call R the distance from O to any point Q in the aperture. Then the distance $PQS = r + r'$ can be expressed in terms of R as follows:

$$r + r' = (h^2 + R^2)^{1/2} + (h'^2 + R^2)^{1/2}$$

$$= h + h' + \frac{1}{2} R^2 \left(\frac{1}{h} + \frac{1}{h'} \right) + \cdots \tag{5.35}$$

where h and h' are the distances OP and OS, respectively. Now suppose that the aperture is divided up into regions bounded by concentric circles, $R = constant$, defined such that $r + r'$ differs by $\frac{1}{2}$ wavelength from one boundary to the next. These regions are called *Fresnel zones*. From (5.35) the successive radii are $R_1 = \sqrt{\lambda L}$, $R_2 = \sqrt{2\lambda L}$, . . . $R_n = \sqrt{n\lambda L}$, where λ is the wavelength, and

$$L = \left(\frac{1}{h} + \frac{1}{h'} \right)^{-1} \tag{5.36}$$

If R_n and R_{n+1} are the inner and outer radii of the $n + 1$st zone, then the area is $\pi R_{n+1}^2 - \pi R_n^2 = \pi R_1^2$. This is independent of n. The areas of the complete zones are therefore all equal.

Typically, the radii of the low-order Fresnel zones are very small.

For example, if $h = h' = 50$ cm and $\lambda = 600$ nm, then we find $R_1 = (\lambda L)^{1/2} = 0.4$ mm, approximately. Also, since R_n is proportional to $n^{1/2}$, we see that the radius of the hundredth zone is only about 4 mm.

The optical disturbance at P can be evaluated in terms of the contributions from the various Fresnel zones, U_1, U_2, U_3, ... Since the mean phase changes by exactly 180 degrees from one zone to the next, the sum of the contributions to the amplitude $|U_p|$ can be expressed as

$$|U_p| = |U_1| - |U_2| + |U_3| - \cdots \qquad (5.37)$$

Consider, for example, the case of a circular aperture centered at O. If the aperture includes precisely n complete zones, then, since the areas are equal, the $|U|$s are all approximately the same. Hence the sum will be very nearly zero if n is even, and approximately the value of $|U_1|$ alone if n is odd.

Consideration of the obliquity factor and the radial distance factor in the Fresnel-Kirchhoff formula [Equation (5.11)] shows that the value of $|U_n|$ decreases slowly with increasing n. As a result, as $n \rightarrow \infty$ the total optical disturbance at P for the case of an infinitely large aperture, that is, no aperture at all, is approximately one half the contribution from the first Fresnel zone alone. To show this (at least qualitatively) we group the terms in Equation (5.37) in the following way:

$$|U_p| = \tfrac{1}{2}|U_1| + (\tfrac{1}{2}|U_1| - |U_2| + \tfrac{1}{2}|U_3|) + (\tfrac{1}{2}|U_3| - |U_4| + \tfrac{1}{2}|U_5|) + \cdots \qquad (5.38)$$

If the decrease with increasing n is very slow, the value of any $|U_n|$ is approximately equal to the mean value of the two adjacent $|U|$s, so that the terms in parentheses very nearly cancel. Thus $\tfrac{1}{2}|U_1|$ is the optical disturbance at P when there is no aperture at all.

Suppose we have a circular obstacle instead of an aperture. The construction of the Fresnel zones is now started at the edge of the obstacle. The value of $|U_p|$ is then, as above, just half the contribution from the first unobstructed zone. As a result the center of the shadow of a circular opaque object shows a bright spot.[2] The irradiance at the bright spot is very nearly the same as it would be in the absence of the obstacle.

In the case of an irregular obstacle or aperture, the appearance of the Fresnel zones as seen from the receiving point P is shown in Figure 5.21. In the illuminated region (a) the outer zones are partially blocked. Thus the higher terms in Equation (5.37) diminish more rap-

[2] The existence of such a bright spot was first predicted by Poisson in 1818 and later confirmed experimentally by Arago and Fresnel.

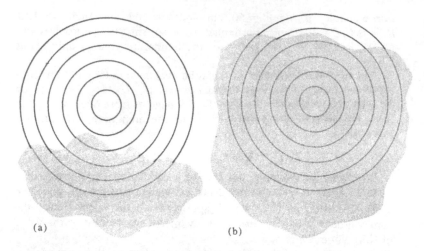

(a)   (b)

Figure 5.21.   Fresnel zones of a point source behind an irregular obstacle. (a) Outside geometrical shadow; (b) inside geometrical shadow.

idly than if there were no obstacle, but the beginning terms are unaffected. As a result the value of $|U_p|$ is hardly changed. On the other hand, in (b) the central zones are completely blocked and the outer zones are partially obstructed. Accordingly, the terms in the summation diminish at both ends and the result is almost complete cancellation. Thus if P is in the illuminated region, the presence of the obstacle makes little or no difference; but if it is in the shadow region, the optical disturbance is very small, which is roughly in agreement with geometrical optics. Diffraction fringes appear around the shadow only if the irregularities at the edge of the obstacle are small compared to the radius of the first Fresnel zone.

Zone Plate   If an aperture is constructed so as to obstruct alternate Fresnel zones, say the even-numbered ones, then the remaining terms in the summation are all of the same sign. Thus

$$|U_p| = |U_1| + |U_3| + |U_5| + \cdots \qquad (5.39)$$

Such an aperture is called a *zone plate*. It acts very much like a lens, because $|U_p|$, and hence the irradiance at P, is now much larger than if there were no aperture. The equivalent focal length is L in (5.36). It is given by

$$L = \frac{R_1{}^2}{\lambda} \qquad (5.40)$$

Zone plates can be made by photographing a drawing similar to that of Figure 5.22. The resulting photographic transparency can

Figure 5.22. A zone plate.

focus light and form images of distant objects. It is a very chromatic lens, however, since the focal length is inversely proportional to the wavelength.

Rectangular Aperture Fresnel diffraction by an aperture of rectangular shape is treated by using the Fresnel-Kirchhoff formula [Equation (5.11)]. We shall employ Cartesian coordinates x,y in the aperture plane as shown in Figure 5.23. Then $R^2 = x^2 + y^2$, and therefore,

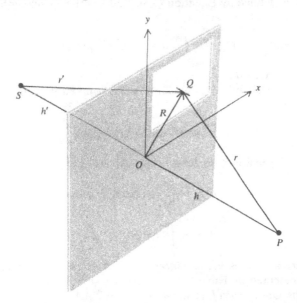

Figure 5.23. Geometry of a rectangular aperture.

referring to Equations (5.35) and (5.36) we have approximately

$$r + r' = h + h' + \frac{1}{2L}(x^2 + y^2) \qquad (5.41)$$

Again, as in the treatment of Fraunhofer diffraction, we shall assume that the obliquity factor cos $(\mathbf{n,r})$ − cos $(\mathbf{n,r'})$ and the radial factor $1/rr'$ vary so slowly compared to the exponential factor $e^{ik(r+r')}$ that they can be taken outside the integrand. The Fresnel-Kirchhoff formula then becomes

$$U_p = C \int_{x_1}^{x_2} \int_{y_1}^{y_2} e^{ik(x^2+y^2)/2L}\, dx\, dy$$

$$= C \int_{x_1}^{x_2} e^{ikx^2/2L}\, dx \int_{y_1}^{y_2} e^{iky^2/2L}\, dy \qquad (5.42)$$

where C includes all other factors. Upon introducing the dimensionless variables u and v defined as

$$u = x \sqrt{\frac{k}{\pi L}} \qquad v = y \sqrt{\frac{k}{\pi L}}$$

or equivalently

$$u = x \sqrt{\frac{2}{\lambda L}} \qquad v = y \sqrt{\frac{2}{\lambda L}} \qquad (5.43)$$

where L is defined by Equation (5.36) and λ is the wavelength, we can write

$$U_p = U_1 \int_{u_1}^{u_2} e^{i\pi u^2/2}\, du \int_{v_1}^{v_2} e^{i\pi v^2/2}\, dv \qquad (5.44)$$

where $U_1 = C\pi L/k$.

The integrals in Equation (5.44) are evaluated in terms of the integral

$$\int_0^s e^{i\pi w^2/2}\, dw = C(s) + iS(s) \qquad (5.45)$$

in which the real and the imaginary parts are given by

$$C(s) = \int_0^s \cos(\pi w^2/2)\, dw$$

$$S(s) = \int_0^s \sin(\pi w^2/2)\, dw \qquad (5.46)$$

These are known as *Fresnel integrals*. A short table of numerical values is presented in Table 5.2, and a graph showing $C(s)$ versus $S(s)$, called the *Cornu spiral*, is shown in Figure 5.24.

Table 5.2. FRESNEL INTEGRALS

s	$C(s)$	$S(s)$
0.0	0.000	0.000
0.2	0.200	0.004
0.4	0.398	0.033
0.6	0.581	0.111
0.8	0.723	0.249
1.0	0.780	0.438
1.2	0.715	0.623
1.4	0.543	0.714
1.6	0.366	0.638
1.8	0.334	0.451
2.0	0.488	0.343
2.5	0.457	0.619
3.0	0.606	0.496
3.5	0.533	0.415
4.0	0.498	0.420
∞	0.500	0.500

The Cornu spiral is useful for graphical evaluation of the Fresnel integrals. The limit points s_1 and s_2 are marked on the spiral. A straight line segment drawn from s_1 to s_2 [Figure 5.24(b)] then gives the value of the integral $\int_{s_1}^{s_2} e^{i\pi w^2/2}\, dw$. The length of the line segment is the magnitude of the integral, and the projections on the C and S axes are the real and imaginary parts, respectively. Also, from Equations (5.46) we see that $(dC)^2 + (dS)^2 = (ds)^2$, hence ds represents an element of arc. The total arc length on the Cornu spiral is equal to the difference between the two limits, namely, $s_2 - s_1$. This difference is proportional to the size of the aperture, that is,

$$s_2 - s_1 = u_2 - u_1 = (x_2 - x_1)\sqrt{\frac{2}{\lambda L}}$$

for the x dimension, and

$$s_2 - s_1 = v_2 - v_1 = (y_2 - y_1)\sqrt{\frac{2}{\lambda L}}$$

for the y dimension.

The limiting case of an infinite aperture, that is, *no diffraction screen at all*, is obtained by setting $u_1 = v_1 = -\infty$ and $u_2 = v_2 = +\infty$. Since $C(\infty) = S(\infty) = \frac{1}{2}$, and $C(-\infty) = S(-\infty) = -\frac{1}{2}$, we obtain the value $U_1(1 + i)^2$ for the unobstructed optical disturbance. On the Cornu

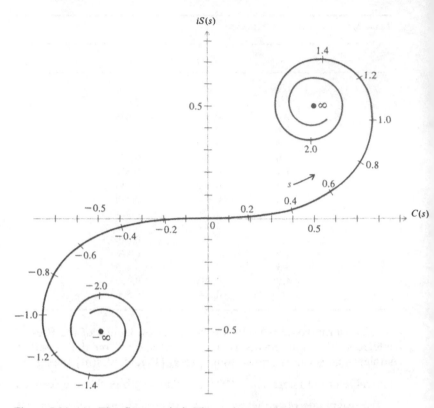

Figure 5.24. (a)   The Cornu spiral. The scale of w is marked on the curve.

spiral this would be U_1 times the length of the line from $-\infty$ to ∞ [Figure 5.24(b)]. Setting this equal to U_0, we can express the general case in the normalized form

$$U_\mathrm{p} = \frac{U_0}{(1+i)^2} \left[C(u) + iS(u) \right]_{u_1}^{u_2} \left[C(\mathrm{v}) + iS(\mathrm{v}) \right]_{\mathrm{v}_1}^{\mathrm{v}_2} \qquad (5.47)$$

Strictly speaking, very large values of the parameters u, v, or s would be inconsistent with the approximation expressed by Equation (5.41). However, in normal cases of interest most of the contribution to U_p comes from the lower-order Fresnel zones in the aperture, corresponding to low values of the above parameters, hence the approximation is still valid.

Slit and Straightedge   Fresnel diffraction by a long slit is treated as a limiting case of a rectangular aperture, namely, by letting $u_1 = -\infty$ and $u_2 = +\infty$ in Equation (5.47). This yields the formula

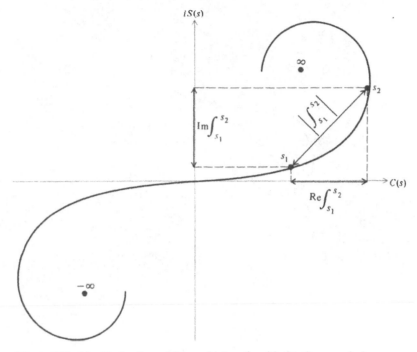

Figure 5.24. (b)   Evaluation of Fresnel integrals with the Cornu spiral.

$$U_p = \frac{U_0}{1+i} \left[C(v) + iS(v) \right]_{v_1}^{v_2} \tag{5.48}$$

for the slit where v_1 and v_2 define the slit edges.

The straightedge is similarly taken as a limiting case of a slit: $v_1 = -\infty$. This gives

$$U_p = \frac{U_0}{1+i} \left[C(v) + iS(v) \right]_{-\infty}^{v_2}$$

$$= \frac{U_0}{1+i} \left[C(v_2) + iS(v_2) + \frac{1}{2} + \frac{1}{2}i \right] \tag{5.49}$$

which is a function of only the one variable v_2. This variable specifies the position of the diffracting edge. If the receiving point P is exactly at the geometrical shadow edge, then $v_2 = 0$. We have then $U_p = [U_0/(1 + i)] \, (\frac{1}{2} + \frac{1}{2}i) = \frac{1}{2}U_0$. Hence the amplitude at the shadow edge is one half, and the irradiance is one fourth the unobstructed value. A plot of $I_p = |U_p|^2$ as given by Equation (5.49) is shown in Figure 5.25. Here I_p is plotted as a function of v_2. This is equivalent to having a fixed position for the receiving point and varying the posi-

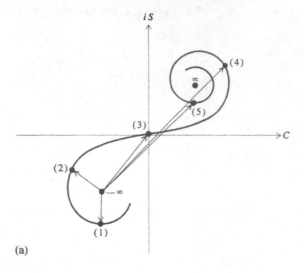

(a)

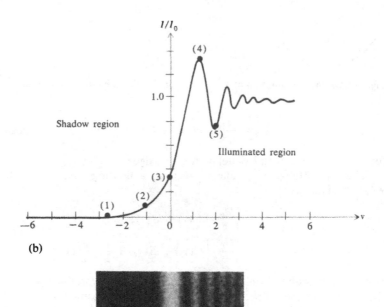

(b)

Figure 5.25. Fresnel diffraction by a straightedge. (a) Points on the Cornu spiral; (b) corresponding points on the intensity curve; $v = 0$ defines the geometrical shadow edge. A photograph of the diffraction pattern is shown below.

tion of the diffracting edge. The result is virtually the same as a diffraction pattern. From the graph it can be seen that the irradiance falls off rapidly and monotonically in the shadow zone ($v_2 < 0$) as $v_2 \rightarrow -\infty$. On the other hand, in the illuminated zone ($v_2 > 0$) the irradiance oscillates with diminishing amplitude about the unobstructed value U_0 as $v_2 \rightarrow +\infty$. The highest irradiance occurs just inside the illuminated region at the point $v_2 \approx 1.25$, where I_p is 1.37 times the irradiance of the unobstructed wave. This is seen as a bright fringe next to the geometrical shadow.

5.6 Applications of the Fourier Transform to Diffraction

Let us return to the discussion of Fraunhofer diffraction. We now consider the general problem of diffraction by an aperture having not only an arbitrary shape, but also an arbitrary transmission including phase retardation, which may vary over different parts of the aperture.

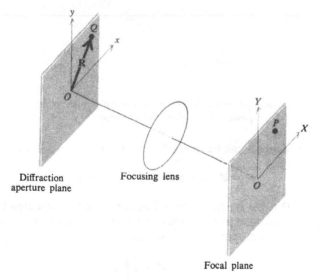

Figure 5.26. Geometry of the general diffraction problem.

We choose coordinates as indicated in Figure 5.26. The diffracting aperture lies in the xy plane, and the diffraction pattern appears in the XY plane, which is the focal plane of the focusing lens. According to elementary geometrical optics, all rays leaving the diffracting aperture in a given direction, specified by direction cosines α, β, and γ, are brought to a common focus. This focus is located at the point $P(X,Y)$ where $X \approx L\alpha$ and $Y \approx L\beta$, L being the focal length of the lens. The assumption is made here that α and β are small, so that $\alpha \approx \tan \alpha$ and $\beta \approx \tan \beta$. We also assume that $\gamma \approx 1$.

Now the path difference δr, between a ray starting from the point $Q(x,y)$ and a parallel ray starting from the origin O, is given by $\mathbf{R} \cdot \hat{\mathbf{n}}$, (Figure 5.27), where $\mathbf{R} = \hat{\mathbf{i}}x + \hat{\mathbf{j}}y$ and $\hat{\mathbf{n}}$ is a unit vector in the direction

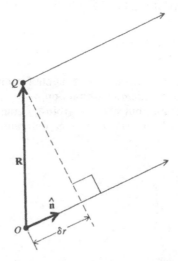

Figure 5.27. Path difference between two parallel rays of light originating from points O and Q in the xy plane.

of the ray. Since $\hat{\mathbf{n}}$ can be expressed as $\hat{\mathbf{n}} = \hat{\mathbf{i}}\alpha + \hat{\mathbf{j}}\beta + \hat{\mathbf{k}}\gamma$, then

$$\delta r = \mathbf{R} \cdot \hat{\mathbf{n}} = x\alpha + y\beta = x \frac{X}{L} + y \frac{Y}{L} \tag{5.50}$$

It follows that the fundamental diffraction integral [Equation (5.16)] giving the diffraction pattern in the XY plane is, aside from a constant multiplying factor, expressible in the form

$$U(X,Y) = \iint e^{ik\delta r} \, d\mathscr{A} = \iint e^{ik(xX+yY)/L} \, dx \, dy \tag{5.51}$$

This is the case for a uniform aperture.

For a uniform rectangular aperture the double integral reduces to the product of two one-dimensional integrals. The result is stated earlier in Section 5.4.

For a nonuniform aperture we introduce a function $g(x,y)$ called the *aperture function*. This function is defined such that $g(x,y) \, dx \, dy$

is the amplitude of the diffracted wave originating from the element of area $dx\,dy$. Thus instead of Equation (5.51), we have the more general formula

$$U(X,Y) = \iint g(x,y)\ e^{i(Xx + Yy)}\ dx\,dy \qquad (5.52)$$

It is convenient at this point to introduce the quantities

$$\mu = \frac{kX}{L} \quad \text{and} \quad \nu = \frac{kY}{L} \qquad (5.53)$$

μ and ν are called *spatial frequencies*, although they have the dimensions of reciprocal length, that is, wavenumber. We now write Equation (5.52) as

$$U(\mu,\nu) = \iint g(x,y)\ e^{i(\mu x + \nu y)}\ dx\,dy \qquad (5.54)$$

We see that the functions $U(\mu,\nu)$ and $g(x,y)$ constitute a two-dimensional Fourier transform pair. The diffraction pattern, in this context, is actually a Fourier resolution of the aperture function.

Consider as an example a grating. For simplicity we treat it as a one-dimensional problem. The aperture function $g(y)$ is then a periodic step function as shown in Figure 5.28. It is represented by a Fourier series of the form

$$g(y) = g_0 + g_1 \cos(\nu_0 y) + g_2 \cos(2\nu_0 y) + \cdots \qquad (5.55)$$

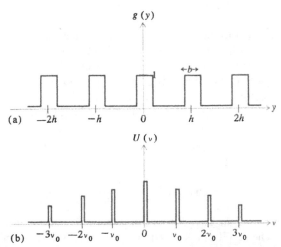

Figure 5.28.   Aperture function for a grating and its Fourier transform.

The fundamental spatial frequency ν_0 is given by the periodicity of the grating, namely,

$$\nu_0 = \frac{2\pi}{h} \tag{5.56}$$

where h is the grating spacing. This dominant spatial frequency appears in the diffraction pattern as the first-order maximum, the amplitude of which is proportional to g_1. Maxima of higher order correspond to higher Fourier components of the aperture function $g(y)$. Thus if the aperture function were of the form of a cosine function $g_0 + g_1 \cos(\nu_0 y)$ instead of a periodic step function, then the diffraction pattern would consist only of the central maximum and the two first-order maxima. Second or higher diffraction orders would not appear.

Apodization   Apodization (literally "to remove the feet") is the name given to any process by which the aperture function is altered in such a way as to produce a redistribution of energy in the diffraction pattern. Apodization is usually employed to reduce the intensity of the secondary diffraction maxima.

It is perhaps easiest to explain the theory of apodization by means of a specific example. Let the aperture consist of a single slit. The aperture function in this case is a single step function: $g(y) = 1$ for $-b/2 < y < b/2$ and $g(y) = 0$ otherwise (Figure 5.29). The corresponding diffraction pattern, expressed in terms of spatial frequencies, is

$$U(\nu) = \int_{-b/2}^{+b/2} e^{i\nu y}\, dy = b\, \frac{\sin\left(\tfrac{1}{2}\nu b\right)}{\left(\tfrac{1}{2}\nu b\right)} \tag{5.57}$$

This is equivalent to the normal case already discussed in Section 5.5.

Suppose now that the aperture function is altered by apodizing in such a way that the resultant aperture transmission is a cosine function: $g(y) = \cos(\pi y/b)$ for $-b/2 < y < b/2$ and zero otherwise, as shown in Figure 5.29. This could be accomplished, for example, by means of a suitably coated-glass plate placed over the aperture. The new diffraction pattern is given by

$$U(\nu) = \int_{-b/2}^{+b/2} \cos\left(\frac{\pi y}{b}\right) e^{i\nu y}\, dy$$

$$= \cos(\nu b/2)\left(\frac{1}{\nu - \pi/b} - \frac{1}{\nu + \pi/b}\right) \tag{5.58}$$

A comparison of the two diffraction patterns is shown graphically in the figure. The result of apodization in this case is a substantial reduction in the secondary maxima relative to the central maximum; in

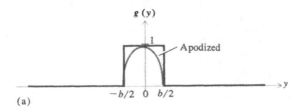

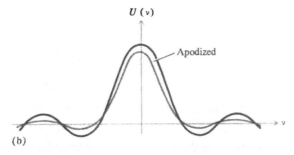

Figure 5.29. (a) Aperture functions for a slit and an apodized slit; (b) the Fourier transforms.

other words, apodization has suppressed the higher spatial frequencies.

In a similar way it is possible to apodize the circular aperture of a telescope so as to reduce greatly the relative intensities of the diffraction rings that appear around the images of stars (discussed in Section 5.5). This enhances the ability of the telescope to resolve the image of a dim star near that of a bright one.

Spatial Filtering Consider the diagram shown in Figure 5.30. Here the xy plane represents the location of some *coherently* illuminated object.[3] This object is imaged by an optical system (not shown), the image appearing in the $x'y'$ plane. The diffraction pattern $U(\mu,\nu)$ of the object function $g(x,y)$ appears in the $\mu\nu$ plane. This plane is analogous to the XY plane in Figure 5.26. Hence, from Equation (5.54) $U(\mu,\nu)$ is the Fourier transform of $g(x,y)$. The image function $g'(x',y')$ that appears in the $x'y'$ plane is, in turn, the Fourier transform of $U(\mu,\nu)$. Now if *all* spatial frequencies in the range $\mu = \pm\infty$, $\nu = \pm\infty$ were transmitted equally by the optical system, then, from the properties of the Fourier transform, the image function $g'(x'y')$

[3] For a discussion of the theory of spatial filtering with incoherent illumination see Reference [10].

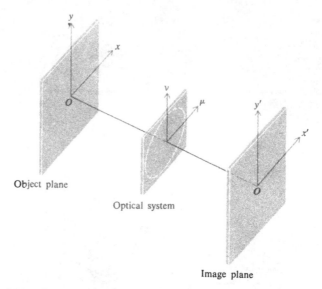

Object plane

Optical system

Image plane

Figure 5.30.   Geometry for the general problem of image formation by an optical system.

would be *exactly* proportional to the object function $g(x,y)$; that is, the image would be a true reproduction of the object. However, the finite size of the aperture at the $\mu\nu$ plane limits the spatial frequencies that are transmitted by the optical system. Furthermore there may be lens defects, aberrations, and so forth, which result in a modification of the function $U(\mu,\nu)$. All of these effects can be incorporated into one function $T(\mu,\nu)$ called the *transfer function* of the optical system. This function is defined implicitly by the equation

$$U'(\mu,\nu) = T(\mu,\nu)\, U(\mu,\nu)$$

Thus

$$g'(x',y') = \int_{-\infty}^{+\infty} \int_{-\infty}^{+\infty} T(\mu,\nu) U(\mu,\nu)\, e^{-i(\mu x' + \nu y')}\, d\mu\, d\nu \qquad (5.59)$$

that is, the image function is the Fourier transform of the product $T(\mu,\nu) \cdot U(\mu,\nu)$. The limits of integration are $\pm\infty$ in a formal sense only. The actual limits are given by the particular form of the transfer function $T(\mu,\nu)$.

The transfer function can be modified by placing various screens and apertures in the $\mu\nu$ plane. This is known as *spatial filtering*. The situation is quite analogous to the filtering of an electrical signal by means of a passive electrical network. The object function is the input signal, and the image function is the output signal. The optical system acts like a filter that allows certain spatial frequencies to be transmitted but rejects others.

Suppose, for example, that the object is a grating so that the object function is a periodic step function. This case can be treated as a one-dimensional problem. The object function $f(y)$ and its Fourier transform $U(\nu)$ are then just those shown in Figure 5.28. Now let the aperture in the $\mu\nu$ plane be such that only those spatial frequencies that lie between $-\nu_{max}$ and $+\nu_{max}$ are transmitted. This means that we have low-pass filtering. From Equation (5.53) we have $\nu_{max} = kb/f$, where $2b$ is the physical width of the aperture in the $\mu\nu$ plane. The transfer function for this case is a step function: $T(\nu) = 1$, $-\nu_{max} < \nu < +\nu_{max}$, and zero otherwise. The image function is, accordingly,

$$g'(y') = \int_{-\nu_{max}}^{+\nu_{max}} U(\nu)e^{-i\nu y'}\, d\nu \qquad (5.60)$$

Without going into the details of the calculation of $g'(y')$, we show in Figure 5.31(a) a graphical plot for some arbitrary choice of ν_{max}. Instead of the sharp step function that constitutes the object, the image is rounded at the corners and also shows small periodic variations.

A high-pass optical filter is obtained by placing in the $\mu\nu$ plane a screen that blocks off the central part of the diffraction pattern. This part of the diffraction pattern corresponds to the low frequencies. The approximate form of the resulting image function is shown in Figure 5.31(b). Only the edges of the grating steps are now visible in the image plane. *The edge detail comes from the higher spatial frequencies.*

A practical example of spatial filtering is the *pinhole spatial filter* which is used in laser work to reduce the spurious fringe pattern that always occurs in the output beam of a helium–neon laser. The beam is brought to a sharp focus by means of a short-focal-length lens. A fine pinhole placed at the focal point constitutes the filter, which removes the higher spatial frequencies and hence improves the beam quality of the laser output. A second lens can be used to render the beam parallel.

Phase Contrast and Phase Gratings The method of phase contrast was invented by the Dutch physicist Zernike. It is used to render visible a transparent object whose index of refraction differs slightly from that of a surrounding transparent medium. Phase contrast is particularly useful in microscopy for examination of living organisms, and so forth. In essence, the method consists of the use of a special type of spatial filter.

To simplify the theory of phase contrast, we shall treat the case of a so-called "phase grating" consisting of alternate strips of high- and

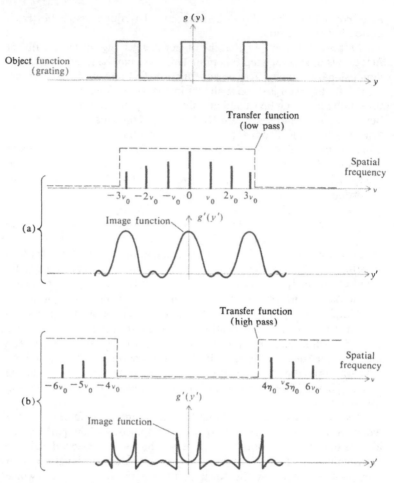

Figure 5.31. Graphs illustrating spatial filtering. (a) Low-pass filtering; (b) high-pass filtering.

low-index material, all strips being perfectly transparent. The grating is coherently illuminated and constitutes the object. The object function is thus represented by the exponential

$$g(y) = e^{i\phi(y)} \tag{5.61}$$

where the phase factor $\phi(y)$ is a periodic step function as shown in Figure 5.32(a). The "height" of the step is the optical-phase difference between the two kinds of strips; that is, $\Delta\phi = kz\,\Delta n$, where z is the thickness and Δn is the difference between the two indices of

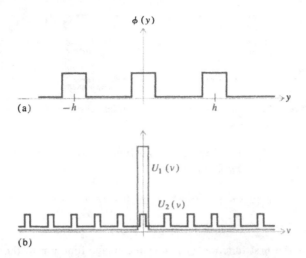

Figure 5.32.   (a) The phase function of a periodic phase grating; (b) Fourier transforms of the aperture U_1 and the grating U_2.

refraction. If we assume that this phase difference is very small, then to a good approximation, we can write

$$g(y) = 1 + i\phi(y) \tag{5.62}$$

The Fourier transform of the above function is

$$U(\nu) = \int_{-\infty}^{\infty} [1 + i\phi(y)]e^{i\nu y}\,dy = \int_{-b/2}^{+b/2} e^{i\nu y}\,dy + i\int_{-b/2}^{+b/2} \phi(y)\,e^{i\nu y}\,dy$$

$$= U_1(\nu) + iU_2(\nu) \tag{5.63}$$

Here $U_1(\nu)$ represents the diffraction pattern of the whole-object aperture. It is essentially zero everywhere except for $\nu \approx 0$; that is, $U_1(\nu)$ contains only very low spatial frequencies. On the other hand, $U_2(\nu)$ represents the diffraction pattern of the periodic step function $\phi(y)$. The two functions are plotted in Figure 5.32(b).

By virtue of the factor i in the result, $U_1 + iU_2$, the two components U_1 and iU_2 are 90 degrees out of phase. The essential trick in the phase-contrast method consists of inserting a spatial filter in the $\mu\nu$ plane, which has the property of shifting the phase of iU_2 by an additional 90 degrees. In practice this is accomplished by means of a device known as a *phase plate*. The physical arrangement is shown in Figure 5.33. The phase plate is just a transparent-glass plate having a small section whose optical thickness is $\frac{1}{4}$ wavelength greater than the remainder of the plate. This thicker section is located in the central part of the $\mu\nu$ plane, that is, in the region of low spatial frequencies. The result of inserting the phase plate is to change the function

Figure 5.33. Physical arrangement of the optical elements for phase contrast microscopy.

$U_1 + iU_2$ to $U_1 + U_2$. The new image function is given by the Fourier transform of the new $U(v)$, namely,

$$g'(y') = \int U_1(v)e^{-ivy'}\, dv + \int U_2(v)e^{-ivy'}\, dv$$

$$= g_1(y') + g_2(y')$$

(5.64)

Now the first function g_1 is just the image function of the whole-object aperture. It represents the constant background. The second function g_2 is the image function for a regular grating of alternate transparent and opaque strips. This means that the phase grating has been rendered visible. It appears in the image plane as alternate bright and dark strips. Although the above analysis has been for a periodic grating, a similar argument can be applied to a transparent-phase object of any shape.

The method of optical-phase contrast has a close analogy in electrical communications. A phase-modulated signal is converted into an amplitude-modulated signal by introduction of a phase shift of 90 degrees to the carrier frequency. This is essentially what the phase plate does in the phase-contrast method. The net result is that phase modulation in the object is converted into amplitude modulation in the image.

5.7 Reconstruction of the Wave Front by Diffraction. Holography

An unusual and interesting method of producing an image — known as the method of *wave-front reconstruction* — has recently become of importance in the field of optics. Although the basic idea was originally proposed by Gabor in 1947 [12], it attracted little attention until the highly coherent light of the laser became available.

In this method a special diffraction screen, called a *hologram,* is used to reconstruct in detail the wave field emitted by the subject. To make the hologram the output from a laser is separated into two beams, one of which illuminates the subject. The other beam, called the *reference beam,* is reflected onto a fine-grained photographic film

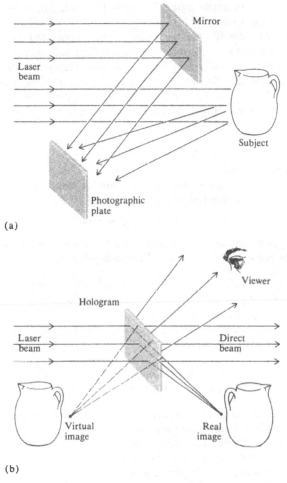

Figure 5.34. (a) Arrangement for producing a hologram; (b) use of the hologram in producing the real and virtual images.

by means of a mirror. The film is exposed simultaneously to the reference beam and the reflected laser light from the subject [Figure 5.34(a)]. The resulting complicated interference pattern recorded by the film constitutes the hologram. It contains all the information needed to reproduce the wave field of the subject.

In use the developed hologram is illuminated with a single beam from a laser as shown in Figure 5.34(b). Part of the resulting diffracted wave field is a precise, three-dimensional copy of the original wave reflected by the subject. The viewer looking at the hologram sees the image in depth and by moving his head can change his perspective of the view.

In order to simplify the discussion of the theory of holography, we shall assume that the reference beam is collimated; that is, it consists of plane waves, although this is not actually necessary in practice. Let x and y be the coordinates in the plane of the recording photographic plate, and let $U(x,y)$ denote the complex amplitude of the reflected wave front in the xy plane. Since $U(x,y)$ is a complex number, we can write it as

$$U(x,y) = a(x,y)e^{i\phi(x,y)} \qquad (5.65)$$

where $a(x,y)$ is real.

Similarly, let $U_0(x,y)$ denote the complex amplitude of the reference beam. Since this beam is plane, we can write

$$U_0(x,y) = a_0\, e^{i(\mu x + \nu y)} \qquad (5.66)$$

where a_0 is a constant and μ and ν are the spatial frequencies of the reference beam in the xy plane. They are given by

$$\mu = k \sin \alpha \qquad \nu = k \sin \beta \qquad (5.67)$$

in which k is the wave number of the laser light, and α and β specify the direction of the reference beam.

The irradiance $I(x,y)$ that is recorded by the photographic film is thus given by the expression

$$I(x,y) = \|U + U_0\|^2 = a^2 + a_0^2 + aa_0 e^{i[\phi(x,y)-\mu x - \nu y]} + aa_0 e^{-i[\phi(x,y)-\mu x-\nu y]}$$
$$= a^2 + a_0^2 + 2aa_0 \cos\left[\phi(x,y) - \mu x - \nu y\right] \qquad (5.68)$$

This is actually an interference pattern. It contains information in the form of amplitude and phase modulations of the spatial frequencies of the reference beam. The situation is somewhat analogous to the impression of information on the carrier wave of a radio transmitter by means of amplitude or phase modulation.

When the developed hologram is illuminated with a single beam U_0 similar to the reference beam, the resulting transmitted wave U_T is proportional to U_0 times the transmittance of the hologram at the point (x,y). The transmittance will be proportional to $I(x,y)$. Hence, except for a constant proportionality factor that we ignore,

$$U_T(x,y) = U_0 I = a_0(a^2 + a_0^2)e^{i(\mu x + \nu y)} + a_0^2 a\, e^{i\phi} + a_0^2 a e^{-i(\phi - 2\mu x - 2\nu y)}$$
$$= (a^2 + a_0^2)U_0 + a_0^2 U + a^2 U^{-1} U_0^{-2} \qquad (5.69)$$

The hologram acts somewhat like a diffraction grating. It produces a direct beam and two first-order diffracted beams on either side of the direct beam [Figure 5.34(b)]. The term $(a^2 + a_0^2)U_0$ in Equation

(5.69) comprises the direct beam. The term $a_0^2 U$ represents one of the diffracted beams. Since it is equal to a constant times U, this beam is the one that reproduces the reflected light from the subject and forms the virtual image. The last term represents the other diffracted beam and gives rise to a real image.

We shall not attempt to prove the above statements in detail. They can be verified by considering a very simple case, namely, that in which the subject is a single white line on a dark background. In this case the hologram turns out to be, in fact, a simple periodic grating. The zero order of the diffracted light is the direct beam, whereas the two first orders on either side comprise the virtual and the real images.

In holography the viewer always sees a positive image whether a positive or a negative photographic transparency is used for the hologram. The reason for this is that a negative hologram merely produces a wave field that is shifted 180 degrees in phase with respect to that of a positive hologram. Since the eye is insensitive to this phase difference, the view seen by the observer is identical in the two cases.

Remarkable technical advances have been made in the field of holography in recent years. Holography in full color is possible by using three different laser wavelengths instead of just one, the holographic record being on black-and-white film. The holographic principle has been extended to include the use of acoustic waves for imaging in optically opaque media and to microwaves for long-distance holography

Holographic Interferometry   One of the most notable applications of holography is in the field of interferometry. In this application the surface to be tested can be irregular and diffusely reflecting instead of smooth and highly polished as is required for the ordinary Michelson and Twyman-Green type of interferometric work. In *double-exposure* holographic interferometry two separate exposures are made on a single recording film. If the surface under study undergoes any deformation or movement during the time interval between exposures, such movement is revealed on the reconstructed image in the form of interference fringes. In *double-pulse* holography the two exposures are produced by short, intense laser pulses from a high-power pulsed laser. These pulses are closely spaced in time so that the holographic image fringes can show motion, vibration patterns, and so on. The method is especially useful for nondestructive testing. For more information on the subject of holography, the reader is encouraged to consult a text such as *An Introduction to Coherent Optics and Holography* by G. W. Stroke [38].

PROBLEMS

5.1 In a diffraction experiment a point (pinhole) source of wavelength 600 nm is to be used. The distance from the source to the diffracting aperture is 10 m, and the aperture is a hole of 1-mm diameter. Determine whether Fresnel or Fraunhofer diffraction applies when the screen-to-aperture distance is (a) 1 cm (b) 2 m.

5.2 A collimated beam of light from a helium–neon laser ($\lambda = 633$ nm) falls normally on a slit 0.5 mm wide. A lens of 50 cm focal length placed just behind the slit focuses the diffracted light on a screen located at the focal distance. Calculate the distance from the center of the diffraction pattern (central maximum) to the first minimum and to the first secondary maximum.

5.3 If white light were used in the above diffraction experiment, for what wavelength would the fourth maximum coincide with the third maximum for red light ($\lambda = 650$ nm)?

5.4 In a single-slit diffraction pattern the intensity of the successive bright fringes falls off as we go out from the central maximum. Approximately which fringe number has a peak intensity that is $\frac{1}{2}$ percent of the central fringe intensity? (Assume Fraunhofer diffraction applies.)

5.5 Prove that the secondary maxima of a single-slit Fraunhofer diffraction pattern occur at the points for which $\beta = \tan \beta$. Show that the first three roots are given by $\beta = 1.43\pi$, 2.46π, and 3.47π, approximately. Show further that for large n, the roots approach the values $(n + \frac{1}{2})\pi$. where n is an integer.

5.6 Find the value of I/I_0 for the first diagonal maximum of the Fraunhofer diffraction pattern of a rectangular aperture. (The diagonal maxima are those that occur on the line $\alpha = \beta$.)

5.7 What size telescope (radius of aperture) would be required to resolve the components of a double star whose linear separation is 100 million km and whose distance from the earth is 10 light years? (Take $\lambda = 500$ nm.)

5.8 In the Fraunhofer diffraction pattern of a double slit, it is found that the fourth secondary maximum is missing. What is the ratio of slit width b to slit separation h?

5.9 Show that the Fraunhofer diffraction pattern of a double slit reduces to that of a single slit of width $2b$ when the slit width is equal to the separation, that is, when $h = b$.

5.10 (a) A grating is used to resolve the sodium D lines (589.0 nm and 589.6 nm) in the first order. How many rulings are needed? (b) If the focal length of the focusing lens is 20 cm, and the total width of the grating is 2 cm, what is the linear separation at the focal plane between the two D lines?

5.11 A grating has 100 lines. What is the ratio of the intensity of a primary maximum to that of the first secondary maximum?

5.12 Show that there are $2 + 2h/b$ maxima under the central diffraction envelope of a double-slit pattern, where h is the slit separation and b is the slit width.

5.13 A grating has 1000 lines/mm of width. How wide must the grating be in order to resolve the mode structure of a HeNe laser beam of wavelength 633 nm? The frequency difference between the modes is 450 MHz.

5.14 What is the minimum resolvable wavelength separation for a grating of 1200 lines/mm having a width of 5 cm? The wavelength is 500 nm, and the grating is to be used in the first order.

5.15 A point source $S(\lambda = 500$ nm) is placed 1 m from an aperture consisting of a hole of 1-mm radius in which there is a circular opaque obstacle whose radius is $\frac{1}{2}$ mm, as shown in Figure 5.35.

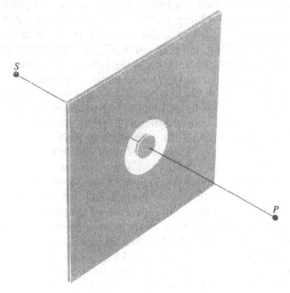

Figure 5.35.   Dimensions of the diffraction aperture for Problem 5.15.

The receiving point P is 1 m from the aperture. What is the irradiance at P compared to the irradiance if the aperture were removed?

5.16 A radiotelescope is observing a distant point source at a wavelength of 20 cm. As the moon passes in front of the source, a

Fresnel diffraction pattern is traced out by the telescope's recorder. What is the time interval between the first maximum and the first minimum? (Assume the edge of the moon to be effectively straight.)

5.17   Apply Equation (5.12) directly to show that the value of U_p, contributed by the first Fresnel zone alone, is twice the value with no aperture at all.

5.18   Find the intensity at the receiving point P in Problem 5.15 if the aperture is an open square 2×2 mm.

5.19   Using Cornu's spiral, make a plot of the Fresnel diffraction pattern of (a) a slit and (b) a complementary opaque strip. Note how Babinet's principle applies here. (Take the equivalent width to be $\Delta v = 3$.) (c) If the actual width is 1 mm and the incident light is parallel, $\lambda = 500$ nm, what is the observing position in order that $\Delta v = 3$?

5.20   An object consists of a single white strip of width b. Treating the case as a one-dimensional problem, find the spatial frequency function $U(v)$ for coherent illumination of the object.

5.21   Referring to Problem 5.20, if the μv aperture is limited to $\pm v_{max}$, where v_{max} lies at the second zero of the $U(v)$ function, find the resulting image function $g'(y')$. Express as an integral.

5.22   A simple hologram is made in the following manner:

The object is a single narrow white strip located a distance d from the base of the recording plate. The wavelength of the laser is λ. The plate is illuminated normally by the reference beam. Show that the resulting pattern on the hologram is a one-dimensional grating with a variable spacing in the y direction. Give the numerical values of this spacing for $\lambda = 6328$ Å, $d = 10$ cm, when $y = 0$, 1 cm, 5 cm, 10 cm.

5.23   Referring to Problem 5.22, show in detail how, if the hologram is illuminated with monochromatic light, there will be two diffracted beams, one producing a real image of the strip, the other producing a virtual image. One beam appears to diverge from a line 0 corresponding to the original object, the other converges to a line 0', the real image. Find the actual angles of diffraction for various values of y given in Problem 5.22. Will there be a second-order (or higher-order) diffracted beam?

5.24   Calculate the diffraction pattern of an apodized slit for which the transmission function is $[\frac{1}{2} + \frac{1}{2} \cos (2\pi y/b)] = g(y)$ for $-b/2 < y < b/2$ and zero otherwise. Find the relative intensity of the first secondary maximum.

CHAPTER 6
Optics of Solids

6.1 General Remarks

The study of the propagation of light through matter, particularly solid matter, comprises one of the important and interesting branches of optics. The many and varied optical phenomena exhibited by solids include such things as selective absorption, dispersion, double refraction, polarization effects, and electro-optical and magneto-optical effects. Many of the optical properties of solids can be understood on the basis of classical electromagnetic theory. The present chapter applies the macroscopic Maxwell theory to the propagation of light through solids. The microscopic origin of the optical properties of solids will be treated classically, since the quantum-theoretical treatment is beyond the scope of this book. But the way in which the phenomena are described by the classical theory gives considerable physical insight and helps to provide a fundamental background for later study.

6.2 Macroscopic Fields and Maxwell's Equations

The electromagnetic state of matter at a given point is described by four quantities:

(1) The volume density of electric charge ρ
(2) The volume density of electric dipoles, called the *polarization* **P**
(3) The volume density of magnetic dipoles, called the *magnetization* **M**
(4) The electric current per unit area, called the *current density* **J**

All of these quantities are considered to be macroscopically averaged in order to smooth out the microscopic variations due to the atomic makeup of all matter. They are related to the macroscopically averaged fields **E** and **H** by the following Maxwell equations:

$$\nabla \times \mathbf{E} = -\mu_0 \frac{\partial \mathbf{H}}{\partial t} - \mu_0 \frac{\partial \mathbf{M}}{\partial t} \qquad (6.1)$$

152

$$\nabla \times \mathbf{H} = \epsilon_0 \frac{\partial \mathbf{E}}{\partial t} + \frac{\partial \mathbf{P}}{\partial t} + \mathbf{J} \qquad (6.2)$$

$$\nabla \cdot \mathbf{E} = -\frac{1}{\epsilon_0} \nabla \cdot \mathbf{P} + \frac{\rho}{\epsilon_0} \qquad (6.3)$$

$$\nabla \cdot \mathbf{H} = -\nabla \cdot \mathbf{M} \qquad (6.4)$$

If one introduces the abbreviation $\mathbf{D}$ for the quantity $\epsilon_0\mathbf{E} + \mathbf{P}$, known as the *electric displacement,* and the abbreviation $\mathbf{B}$ for $\mu_0(\mathbf{H} + \mathbf{M})$, called the *magnetic induction,* then Maxwell's equations assume the more compact forms:

$$\nabla \times \mathbf{E} = -\frac{\partial \mathbf{B}}{\partial t} \qquad (6.5)$$

$$\nabla \times \mathbf{H} = \frac{\partial \mathbf{D}}{\partial t} + \mathbf{J} \qquad (6.6)$$

$$\nabla \cdot \mathbf{D} = \rho \qquad (6.7)$$

$$\nabla \cdot \mathbf{B} = 0 \qquad (6.8)$$

The response of the conduction electrons to the electric field is given by the current equation (Ohm's law)

$$\mathbf{J} = \sigma\mathbf{E}$$

where σ is the conductivity. The constitutive relation

$$\mathbf{D} = \epsilon\mathbf{E}$$

describes the aggregate response of the bound charges to the electric field. The corresponding magnetic relation is

$$\mathbf{B} = \mu\mathbf{H}$$

An alternate way to express the response of the bound charges is

$$\mathbf{P} = (\epsilon - \epsilon_0)\mathbf{E} = \chi\epsilon_0\mathbf{E} \qquad (6.9)$$

which gives the proportionality between the polarization and the impressed electric field. The proportionality factor

$$\chi = \frac{\epsilon}{\epsilon_0} - 1$$

is known as the *electric susceptibility.* In the study of the optics of matter, χ is a most important parameter.

In the case of isotropic media, for example, glass, χ is a scalar quantity having the same value for any direction of the applied electric field. For nonisotropic media, such as most crystals, the magnitude of the polarization varies with the direction of the applied field and, consequently, χ must be expressed as a tensor. We shall find that the χ tensor of a crystal summarizes most of its optical properties.

6.3 The General Wave Equation

In our study of solid-state optics we shall be concerned only with nonmagnetic, electrically neutral media. Hence M and ρ are both zero. Maxwell's equations, in the form expressed by Equations (6.1) to (6.4), then reduce to the following:

$$\nabla \times \mathbf{E} = -\mu_0 \frac{\partial \mathbf{H}}{\partial t} \tag{6.10}$$

$$\nabla \times \mathbf{H} = \epsilon_0 \frac{\partial \mathbf{E}}{\partial t} + \frac{\partial \mathbf{P}}{\partial t} + \mathbf{J} \tag{6.11}$$

$$\nabla \cdot \mathbf{E} = -\frac{1}{\epsilon_0} \nabla \cdot \mathbf{P} \tag{6.12}$$

$$\nabla \cdot \mathbf{H} = 0 \tag{6.13}$$

The general wave equation for the $\mathbf{E}$ field is obtained by taking the curl of Equation (6.10) and the time derivative of Equation (6.11) and eliminating $\mathbf{H}$. The result is

$$\nabla \times (\nabla \times \mathbf{E}) + \frac{1}{c^2} \frac{\partial^2 \mathbf{E}}{\partial t^2} = -\mu_0 \frac{\partial^2 \mathbf{P}}{\partial t^2} - \mu_0 \frac{\partial \mathbf{J}}{\partial t} \tag{6.14}$$

The two terms on the right-hand side of the above equation are called *source terms*. They stem from the presence of polarization charges and conduction charges, respectively, within the medium. The way in which the propagation of light is affected by the sources is revealed by the solution of the wave equation when the source terms are included. In the case of nonconducting media the polarization term $-\mu_0 \, \partial^2 \mathbf{P}/\partial t^2$ is of importance. It turns out that this term leads to an explanation of many optical effects, including dispersion, absorption, double refraction, and optical activity to mention only a few. In the case of metals it is the conduction term $-\mu_0 \, \partial \mathbf{J}/\partial t$ that is important, and the resulting solutions of the wave equation explain the large opacity and high reflectance of metals. Both source terms must be taken into account in the case of semiconductors. The result is a rather complicated wave equation and the solutions are somewhat difficult to interpret. Nevertheless, a qualitative description of many of the optical properties of semiconductors is furnished by classical theory. A rigorous treatment of semiconductor optics must await the application of quantum theory.

6.4 Propagation of Light in Isotropic Dielectrics. Dispersion

In a nonconducting, isotropic medium, the electrons are permanently bound to the atoms comprising the medium and there is no preferential direction. This is what is meant by a simple isotropic dielectric such as glass. Suppose that each electron, of charge $-e$, in a dielectric is displaced a distance $\mathbf{r}$ from its equilibrium position. The resulting macroscopic polarization $\mathbf{P}$ of the medium is given by

$$\mathbf{P} = -N e \mathbf{r} \qquad (6.15)$$

where N is the number of electrons per unit volume. If the displacement of the electron is the result of the application of a static electric field $\mathbf{E}$, and if the electron is elastically bound to its equilibrium position with a force constant K, then the force equation is

$$-e\mathbf{E} = K\mathbf{r} \qquad (6.16)$$

The *static* polarization is therefore given by

$$\mathbf{P} = \frac{Ne^2}{K} \mathbf{E} \qquad (6.17)$$

However, if the impressed field $\mathbf{E}$ varies with time, the above equation is incorrect. In order to find the true polarization in this case, we must take the actual motion of the electrons into account. To do this we consider the bound electrons as classical damped harmonic oscillators. The differential equation of motion is

$$m \frac{d^2\mathbf{r}}{dt^2} + m\gamma \frac{d\mathbf{r}}{dt} + K\mathbf{r} = -e\mathbf{E} \qquad (6.18)$$

The term $m\gamma\,(d\mathbf{r}/dt)$ represents a frictional damping force that is proportional to the velocity of the electron, the proportionality constant being written as $m\gamma$.[1]

Now suppose that the applied electric field varies harmonically with time according to the usual factor $e^{-i\omega t}$. Assuming that the motion of the electron has the same harmonic time dependence, we find that Equation (6.18) becomes

$$(-m\omega^2 - i\omega m\gamma + K)\mathbf{r} = -e\mathbf{E} \qquad (6.19)$$

Consequently, the polarization, from Equation (6.15), is given by

$$\mathbf{P} = \frac{Ne^2}{-m\omega^2 - i\omega m\gamma + K} \mathbf{E} \qquad (6.20)$$

---

[1] The magnetic force $e\mathbf{v}\times\mathbf{B}$ is neglected here. For electromagnetic waves, this force is normally much smaller than the electric force $e\mathbf{E}$.

It reduces to the static value, Equation (6.17), when $\omega = 0$. Thus for a given amplitude of the impressed electric field, the amount of polarization varies with frequency. The phase of $\mathbf{P}$, relative to that of the electric field, also depends on the frequency. This is shown by the presence of the imaginary term in the denominator.

A more significant way of writing Equation (6.20) is

$$\mathbf{P} = \frac{Ne^2/m}{\omega_0{}^2 - \omega^2 - i\omega\gamma}\,\mathbf{E} \tag{6.21}$$

in which we have introduced the abbreviation ω_0 given by

$$\omega_0 = \sqrt{\frac{K}{m}} \tag{6.22}$$

This is the *effective resonance frequency* of the bound electrons.

The polarization formula (6.21) is similar to the amplitude formula for a driven harmonic oscillator, as indeed it should be, since it is the displacement of the elastically bound electrons that actually constitutes the polarization. We should therefore expect to find an optical resonance phenomenon of some kind occurring for light frequencies in the neighborhood of the resonance frequency ω_0. As we shall presently see, this resonance phenomenon is manifest as a large change in the index of refraction of the medium and also by a strong absorption of light at or near the resonance frequency.

To show how the polarization affects the propagation of light, we return to the general wave equation (6.14). For a dielectric there is no conduction term. The polarization is given by Equation (6.21). Hence we have

$$\nabla \times (\nabla \times \mathbf{E}) + \frac{1}{c^2}\frac{\partial^2 \mathbf{E}}{\partial t^2} = \frac{-\mu_0 N e^2}{m}\left(\frac{1}{\omega_0{}^2 - \omega^2 - i\gamma\omega}\right)\frac{\partial^2 \mathbf{E}}{\partial t^2} \tag{6.23}$$

Also, from the linear relationship between $\mathbf{P}$ and $\mathbf{E}$, it follows from (6.12) that $\nabla \cdot \mathbf{E} = 0$. Consequently, $\nabla \times (\nabla \times \mathbf{E}) = -\nabla^2\mathbf{E}$, and the above wave equation reduces to the somewhat simpler one

$$\nabla^2\mathbf{E} = \frac{1}{c^2}\left(1 + \frac{Ne^2}{m\epsilon_0}\cdot\frac{1}{\omega_0{}^2 - \omega^2 - i\gamma\omega}\right)\frac{\partial^2 \mathbf{E}}{\partial t^2} \tag{6.24}$$

after rearranging terms and using the relation $1/c^2 = \mu_0\epsilon_0$.

Let us seek a solution of the form

$$\mathbf{E} = \mathbf{E}_0\,e^{i(\mathcal{K}z - \omega t)} \tag{6.25}$$

This trial solution represents what are called *homogeneous* plane harmonic waves. Direct substitution shows that this is a possible solution provided that

$$\mathcal{K}^2 = \frac{\omega^2}{c^2}\left(1 + \frac{Ne^2}{m\epsilon_0}\cdot\frac{1}{\omega_0{}^2 - \omega^2 - i\gamma\omega}\right) \tag{6.26}$$

The presence of the imaginary term in the denominator implies that the wavenumber $\mathscr{K}$ must be a complex number. Let us inquire as to the physical significance of this. We express $\mathscr{K}$ in terms of its real and imaginary parts as

$$\mathscr{K} = k + i\alpha \qquad (6.27)$$

This amounts to the same thing as introducing a complex index of refraction

$$\mathscr{N} = n + i\kappa \qquad (6.28)$$

where

$$\mathscr{K} = \frac{\omega}{c}\,\mathscr{N} \qquad (6.29)$$

Our solution in Equation (6.25) can then be written as

$$\mathbf{E} = \mathbf{E}_0 e^{-\alpha z}\, e^{i(kz - \omega t)} \qquad (6.30)$$

The factor $e^{-\alpha z}$ indicates that the amplitude of the wave decreases exponentially with distance. This means that as the wave progresses, the energy of the wave is absorbed by the medium. Since the energy in the wave at a given point is proportional to $|\mathbf{E}|^2$, then the energy varies with distance as $e^{-2\alpha z}$. Hence 2α is the *coefficient of absorption* of the medium. The imaginary part κ of the complex index of refraction is known as the *extinction index*. The two numbers α and κ are related by the equation

$$\alpha = \frac{\omega}{c}\,\kappa \qquad (6.31)$$

The phase factor $e^{i(kz - \omega t)}$ indicates that we have a harmonic wave in which the phase velocity is

$$u = \frac{\omega}{k} = \frac{c}{n} \qquad (6.32)$$

From Equations (6.26) and (6.29) we have

$$\mathscr{N}^2 = (n + i\kappa)^2 = 1 + \frac{Ne^2}{m\epsilon_0}\left(\frac{1}{\omega_0^2 - \omega^2 - i\gamma\omega}\right) \qquad (6.33)$$

Equating real and imaginary parts yields the following equations:

$$n^2 - \kappa^2 = 1 + \frac{Ne^2}{m\epsilon_0}\left(\frac{\omega_0^2 - \omega^2}{(\omega_0^2 - \omega^2)^2 + \gamma^2\omega^2}\right) \qquad (6.34)$$

$$2n\kappa = \frac{Ne^2}{m\epsilon_0}\left(\frac{\gamma\omega}{(\omega_0^2 - \omega^2)^2 + \gamma^2\omega^2}\right) \qquad (6.35)$$

from which the optical parameters n and κ may be found.

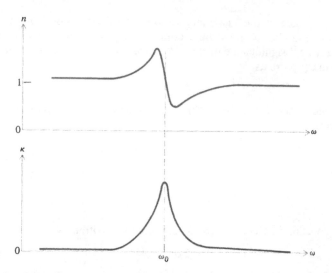

Figure 6.1. Graphs of the index of refraction and extinction coefficient versus frequency near a single resonance line.

Figure 6.1 shows the general way in which n and κ depend on frequency. The absorption is strongest at the resonance frequency ω_0. The index of refraction is greater than unity for small frequencies and increases with frequency as the resonance frequency is approached. This is the case of "normal" dispersion, which is exhibited by most transparent substances over the visible region of the spectrum, the principal resonance frequencies being in the ultraviolet region. At or near the resonance frequency, however, the dispersion becomes "anomalous" in the sense that the index of refraction *decreases* with increasing frequency.

Anomalous dispersion can be observed experimentally if the substance is not too opaque at the resonance frequency. For instance, certain dyes have absorption bands in the visible region of the spectrum and exhibit anomalous dispersion in the region of these bands. Prisms made of these dyes produce a spectrum that is reversed; that is, the longer wavelengths are refracted more than the shorter wavelengths.

Now, in the above discussion it has been tacitly assumed that all of the electrons were identically bound, and hence all had the same resonance frequencies. In order to take into account the fact that different electrons may be bound differently, we may assume that a certain fraction f_1 has an associated resonance frequency ω_1, a fraction f_2 has the resonance frequency ω_2, and so on. The resulting formula for the square of the complex index of refraction is of the form

$$\mathcal{N}^2 = 1 + \frac{Ne^2}{m\epsilon_0} \sum_j \left(\frac{f_j}{\omega_j^2 - \omega^2 - i\gamma_j\omega} \right) \tag{6.36}$$

The summation extends over all the various kinds of electrons indicated by the subscript j. The fractions f_j are known as *oscillator strengths*. The damping constants associated with the various frequencies are denoted by γ_j. Figure 6.2 shows graphically the gen-

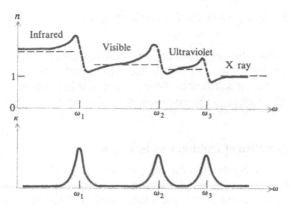

Figure 6.2.   Index of refraction and extinction index for a hypothetical substance with absorption bands in the infrared, visible, and ultraviolet regions of the spectrum.

eral dependence of the real and imaginary parts of $\mathcal{N}$ as determined by Equation (6.36). This graph is intended to show qualitatively the case for a substance, such as glass, which is transparent in the visible region and has absorption bands in the infrared and ultraviolet regions of the spectrum. In the limit of zero frequency, the square of the index approaches the value $1 + (Ne^2/m\epsilon_0) \Sigma f_j/\omega_j^2$. This is just the static dielectric constant of the medium.

In the high-frequency region, the theory predicts that the index should dip below unity and then approach unity from below as ω becomes infinite. This effect is actually seen experimentally. The case of quartz is shown in Figure 6.3. Here the measured index of refraction of quartz is plotted as a function of wavelength for the appropriate region of the spectrum (x-ray region).

If the damping constants γ_j are sufficiently small so that the terms $\gamma_j\omega$ can be neglected in comparison to the quantities $\omega_j^2 - \omega^2$ in Equation (6.36), then the index of refraction is essentially real and its square is given by

$$n^2 = 1 + \frac{Ne^2}{m\epsilon_0} \sum_j \left(\frac{f_j}{\omega_j^2 - \omega^2} \right) \tag{6.37}$$

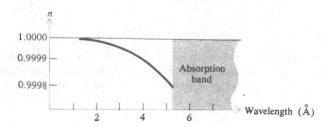

Figure 6.3. Measured index of refraction of quartz in the x-ray region.

It is possible, by an empirical curve-fitting procedure, to make a formula of the above type match the experimental data quite well for many transparent substances. When expressed in terms of wavelength instead of frequency, the equation is known as *Sellmeier's formula*.

6.5 Propagation of Light in Conducting Media

The effects of conduction on the propagation of light through a medium can be treated in much the same manner as the effects of polarization were treated in the preceding section. The difference is that we are now interested in the conduction term in the general wave equation, not the polarization term. Again, owing to the inertia of the conduction electrons, we cannot merely put $J = \sigma E$ for the current density where σ is the static conductivity. We must consider actual motion of the electrons under the action of the alternating electric field of the light wave.

Since the conduction electrons are not bound, there is no elastic restoring force as there was in the case of polarization. The differential equation of motion of the electron is therefore of the form

$$m \frac{dv}{dt} + m\tau^{-1}v = -eE \tag{6.38}$$

where v is the velocity of the electron. The frictional dissipation constant is expressed in the form $m\tau^{-1}$. This constant is related to the static conductivity as we shall presently see. Since the current density is

$$J = -Nev \tag{6.39}$$

where N is now the number of conduction electrons per unit volume, then Equation (6.38) can be expressed in terms of J as follows:

$$\frac{dJ}{dt} + \tau^{-1}J = \frac{Ne^2}{m} E \tag{6.40}$$

The decay of a transient current is governed by the associated homogeneous equation

$$\frac{d\mathbf{J}}{dt} + \tau^{-1}\mathbf{J} = 0 \qquad (6.41)$$

whose solution is $\mathbf{J} = \mathbf{J}_0 e^{-t/\tau}$. Thus a transient current will decay to e^{-1} of its initial value in a time τ. This is called the *relaxation time*. Now for a static electric field, Equation (6.40) becomes

$$\tau^{-1}\mathbf{J} = \frac{Ne^2}{m}\mathbf{E} \qquad (6.42)$$

The static conductivity σ is therefore given by

$$\sigma = \frac{Ne^2}{m}\tau \qquad (6.43)$$

Let us now assume a harmonic time dependence $e^{-i\omega t}$ for both electric field $\mathbf{E}$ and the resulting current $\mathbf{J}$ in our differential Equation (6.40). It follows that

$$(-i\omega + \tau^{-1})\mathbf{J} = \frac{Ne^2}{m}\mathbf{E} = \tau^{-1}\sigma\mathbf{E} \qquad (6.44)$$

Solving for $\mathbf{J}$, we find

$$\mathbf{J} = \frac{\sigma}{1 - i\omega\tau}\mathbf{E} \qquad (6.45)$$

When $\omega = 0$, the above equation reduces to $\mathbf{J} = \sigma\mathbf{E}$, which is the correct equation for the static case.

Using the dynamic expression for $\mathbf{J}$, we find that the general wave Equation (6.14) reduces to

$$\nabla^2\mathbf{E} = \frac{1}{c^2}\frac{\partial^2\mathbf{E}}{\partial t^2} + \frac{\mu_0\sigma}{1 - i\omega\tau}\frac{\partial\mathbf{E}}{\partial t} \qquad (6.46)$$

For a trial solution we take a simple homogeneous plane-wave solution of the type

$$\mathbf{E} = \mathbf{E}_0 \, e^{i(\mathscr{K}z - \omega t)} \qquad (6.47)$$

where, as in Equation (6.26), $\mathscr{K}$ is assumed to be complex. It is easily found that $\mathscr{K}$ must then satisfy the relation

$$\mathscr{K}^2 = \frac{\omega^2}{c^2} + \frac{i\omega\mu_0\sigma}{1 - i\omega\tau} \qquad (6.48)$$

For very low frequencies the above formula reduces to the approximate formula

$$\mathscr{K}^2 \approx i\omega\mu_0\sigma \qquad (6.49)$$

so that $\mathcal{K} \approx \sqrt{i\omega\mu_0\sigma} = (1 + i)\sqrt{\omega\mu_0\sigma/2}$. In this case the real and imaginary parts of $\mathcal{K} = k + i\alpha$ are equal and are given by

$$k \approx \alpha \approx \sqrt{\frac{\omega\sigma\mu_0}{2}} \qquad (6.50)$$

Similarly the real and imaginary parts of $\mathcal{N} = n + i\kappa$ are equal and are given by

$$n \approx \kappa \approx \sqrt{\frac{\sigma}{2\omega\epsilon_0}} \qquad (6.51)$$

The so-called "skin depth" δ of a metal is that distance at which the amplitude of an electromagnetic wave drops to e^{-1} of its value at the surface. Thus

$$\delta = \frac{1}{\alpha} = \sqrt{\frac{2}{\omega\sigma\mu_0}} = \sqrt{\frac{\lambda_0}{c\pi\sigma\mu_0}} \qquad (6.52)$$

where λ_0 is the vacuum wavelength. This shows why good conductors are also highly opaque. A high value of the conductivity σ gives a large coefficient of absorption α and a correspondingly small skin depth. For example, the skin depth in copper ($\sigma = 5.8 \times 10^7$ mho/m) for 1-mm microwaves is about 10^{-4} mm.

Let us return to the more accurate expression for $\mathcal{K}$ given in Equation (6.48). The equivalent form of this equation written in terms of the complex index of refraction, as defined by Equation (6.29), is

$$\mathcal{N}^2 = 1 - \frac{\omega_p{}^2}{\omega^2 + i\omega\tau^{-1}} \qquad (6.53)$$

Here we have introduced *plasma frequency* for the metal. It is defined by the relations

$$\omega_p = \sqrt{\frac{Ne^2}{m\epsilon_0}} = \sqrt{\frac{\mu_0\sigma c^2}{\tau}} \qquad (6.54)$$

By equating real and imaginary parts in Equation (6.53), we find

$$n^2 - \kappa^2 = 1 - \frac{\omega_p{}^2}{\omega^2 + \tau^{-2}} \qquad (6.55)$$

$$2n\kappa = \frac{\omega_p{}^2}{\omega^2 + \tau^{-2}}\left(\frac{1}{\omega\tau}\right) \qquad (6.56)$$

from which the optical "constants" n and κ may be obtained. An explicit algebraic solution of the above pairs of equations is very cumbersome, hence the equations are usually solved numerically for n and κ. According to the above theory, these are determined entirely by the plasma frequency ω_p, the relaxation time τ, and the frequency ω of the light wave.

Typical relaxation times for metals, as deduced from conductivity measurements, are of the order of 10^{-13} s, which corresponds to frequencies in the infrared region of the spectrum. On the other hand plasma frequencies of metals are typically around 10^{15} s^{-1}, corresponding to the visible and near ultraviolet regions. Figure 6.4 shows

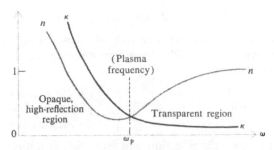

Figure 6.4. Index of refraction and extinction index versus frequency for a metal.

the behavior of n and κ plotted as functions of ω from Equations (6.55) and (6.56). As seen from the figure, the index of refraction n is less than unity for a wide range of frequencies in the region of the plasma frequency. The extinction coefficient κ is very large at low frequencies (long wavelengths). It decreases monotonically with increasing frequency, becoming very small for frequencies greater than the plasma frequency. The metal thus becomes transparent at high frequencies. Qualitative agreement with these predictions of classical theory is obtained in the case of the alkali metals and some of the better conductors such as silver, gold, and copper.

For poor conductors and semiconductors, both free electrons and bound electrons can contribute to the optical properties. Classical theory would, accordingly, yield an equation of the type

$$\mathcal{N}^2 = 1 - \frac{\omega_p{}^2}{\omega^2 + i\omega\tau^{-1}} + \frac{Ne^2}{m\epsilon_0} \sum_j \left(\frac{f_j}{\omega_j{}^2 - \omega^2 - i\gamma_j\omega} \right) \tag{6.57}$$

for the complex index of refraction. It turns out that quantum theory gives a similar relation and, in addition, can predict the values of the various parameters f_j, γ_j, and so forth. The theoretical calculations are difficult, however, as are also the experimental measurements. The optics of semiconductors is one of the most active areas of current experimental and theoretical research.

6.6 Reflection and Refraction at the Boundary of an Absorbing Medium

Let a plane wave be incident on the boundary of a medium having a complex index of refraction

$$\mathcal{N} = n + i\kappa \qquad (6.58)$$

Denote the complex propagation vector of the refracted wave by

$$\mathcal{K} = \mathbf{k} + i\boldsymbol{\alpha} \qquad (6.59)$$

For simplicity we shall consider only the case in which the first medium is nonabsorbing. The following notation will be employed (deleting the amplitudes):

$$
\begin{array}{ll}
e^{i(\mathbf{k}_0\mathbf{r}-\omega t)} & \text{(incident wave)} \\
e^{i(\mathbf{k}_0'\cdot\mathbf{r}-\omega t)} & \text{(reflected wave)} \\
e^{i(\mathcal{K}\cdot\mathbf{r}-\omega t)} = e^{-\boldsymbol{\alpha}\cdot\mathbf{r}}\,e^{i(\mathbf{k}\cdot\mathbf{r}-\omega t)} & \text{(refracted wave)}
\end{array}
$$

As is the case with reflection and refraction at a dielectric interface, discussed earlier in Section 2.6, the requirement that a constant ratio exists among the fields at the boundary plane leads to the equations

$$\mathbf{k}_0 \cdot \mathbf{r} = \mathbf{k}_0' \cdot \mathbf{r} \qquad \text{(at boundary)} \qquad (6.60)$$

$$\mathbf{k}_0 \cdot \mathbf{r} = \mathcal{K} \cdot \mathbf{r} = (\mathbf{k} + i\boldsymbol{\alpha}) \cdot \mathbf{r} \qquad \text{(at boundary)} \qquad (6.61)$$

The first equation gives the usual law of reflection. The second equation, after equating real and imaginary parts, yields

$$\mathbf{k}_0 \cdot \mathbf{r} = \mathbf{k} \cdot \mathbf{r} \qquad (6.62)$$

$$0 = \boldsymbol{\alpha} \cdot \mathbf{r} \qquad (6.63)$$

This result means that, in general, $\mathbf{k}$ and $\boldsymbol{\alpha}$ have different directions. In this case the wave is said to be *inhomogeneous*. In particular, $\boldsymbol{\alpha} \cdot \mathbf{r} = 0$ implies that $\boldsymbol{\alpha}$, which defines the direction of planes of constant amplitude, is always normal to the boundary. On the other hand the planes of constant phase are defined by the vector $\mathbf{k}$, which may have any direction. The situation is illustrated in Figure 6.5. The waves move in the direction of the vector $\mathbf{k}$, but their amplitudes diminish exponentially with the distance from the boundary plane, as shown.

If we denote the angle of incidence by θ and the angle of refraction by ϕ, then Equation (6.62) is equivalent to

$$k_0 \sin \theta = k \sin \phi \qquad (6.64)$$

This equation, which has been derived solely on the basis of phase matching at the boundary, would allow us to find the angle of refrac-

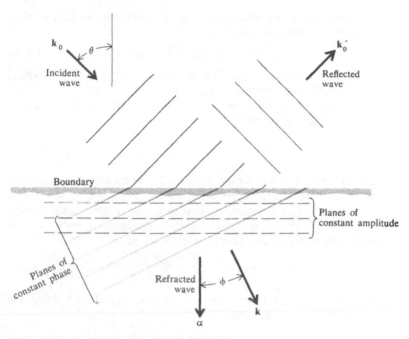

Figure 6.5   Real and imaginary parts of the wave vector in an absorbing medium for the case of oblique incidence of light at the boundary.

tion, given the angle of incidence, if we knew the value of k. As we shall see, however, k is not constant for inhomogeneous waves, but is itself a function of the angle ϕ between the two vectors $\mathbf{k}$ and $\boldsymbol{\alpha}$.

In order to find the required relationship we must go back to the wave equation. This can be written in terms of the complex index of refraction as

$$\nabla^2 \mathbf{E} = \frac{\mathcal{N}^2}{c^2} \frac{\partial^2 \mathbf{E}}{\partial t^2} \tag{6.65}$$

For plane harmonic waves we have $\nabla \to i\mathcal{K}$ and $\partial/\partial t \to -i\omega$, so that

$$\mathcal{K} \cdot \mathcal{K} = \frac{\mathcal{N}^2 \omega^2}{c^2} = \mathcal{N}^2 k_0^2 \tag{6.66}$$

where $k_0 = \omega/c$. Written out in terms of real and imaginary parts, we have

$$(\mathbf{k} + i\boldsymbol{\alpha}) \cdot (\mathbf{k} + i\boldsymbol{\alpha}) = (n + i\kappa)^2 k_0^2 \tag{6.67}$$

By equating real and imaginary parts, we obtain

$$k^2 - \alpha^2 = (n^2 - \kappa^2)k_0^2 \tag{6.68}$$

$$\mathbf{k} \cdot \boldsymbol{\alpha} = k\alpha \cos \phi = n\kappa \, k_0^2 \tag{6.69}$$

After some algebraic manipulation, the above results can be shown to lead to the following formula:

$$k \cos \phi + i\alpha = k_0 \sqrt{\mathcal{N}^2 - \sin^2 \theta} \qquad (6.70)$$

This reduces to $k + i\alpha = k_0 \mathcal{N}$ for normal incidence $(\theta = 0)$ which is the relation for homogeneous waves discussed earlier.

We now express the law of refraction in terms of the complex index of refraction in a purely formal way as

$$\mathcal{N} = \frac{\sin \theta}{\sin \phi} \qquad (6.71)$$

Here the angle ϕ is a complex number. It has no simple physical interpretation, but can be considered as being defined by the above equation. It turns out, however, that ϕ is very useful in simplifying the equations related to reflection and refraction by an absorbing medium. From the above definition of ϕ, we have

$$\cos \phi = \sqrt{1 - \frac{\sin^2 \theta}{\mathcal{N}^2}} \qquad (6.72)$$

This, with Equation (6.70), gives a second formula involving the complex index of refraction:

$$\mathcal{N} = \frac{k \cos \phi + i\alpha}{k_0 \cos \phi} \qquad (6.73)$$

We are now ready to attack the problem of finding the reflectance. We know that the amplitudes of the electric and magnetic fields are related as follows.

$$\mathbf{E}, \mathbf{H} = \frac{1}{\mu_0 \omega} \mathbf{k_0} \times \mathbf{E} \qquad \text{(incident)} \quad (6.74)$$

$$\mathbf{E'}, \mathbf{H'} = \frac{1}{\mu_0 \omega} \mathbf{k_0'} \times \mathbf{E'} \qquad \text{(reflected)} \quad (6.75)$$

$$\mathbf{E''}, \mathbf{H''} = \frac{1}{\mu_0 \omega} \mathcal{K} \times \mathbf{E''} = \frac{1}{\mu_0 \omega} (\mathbf{k} \times \mathbf{E''} + i\alpha \times \mathbf{E''}) \quad \text{(refracted)} \quad (6.76)$$

The coefficient of reflection will be derived for the transverse electric (TE) case. A similar procedure can be used for the transverse magnetic (TM) case. The relevant vectors are essentially the same as those shown in Figure 2.11, with an obvious difference in the **k**s.

The boundary conditions giving the continuity of the tangential components of the electric and magnetic fields for TE polarization are

$$E + E' = E'' \qquad (6.77)$$

$$-H \cos \theta + H' \cos \theta = H''_{\text{tangential}} \qquad (6.78)$$

By applying Equations (6.74) to (6.76) to the second equation, we find

$$-k_0E \cos \theta + k_0E' \cos \theta = -(kE'' \cos \phi + i\alpha E'')$$
$$= -\mathcal{N} k_0 E'' \cos \phi \qquad (6.79)$$

The last step follows from Equation (6.73). We now eliminate E'' from Equations (6.79) and (6.77) to obtain the final result

$$r_s = \frac{\cos \theta - \mathcal{N} \cos \phi}{\cos \theta + \mathcal{N} \cos \phi} \qquad (TE \ polarization) \qquad (6.80)$$

This equation for the ratio of the reflected amplitude to the incident amplitude is of the same form as that for the dielectric case of Equation (2.54). The only difference is that $\mathcal{N}$ and ϕ are now complex. The corresponding equation for TM polarization also turns out to be of the same form as that of the dielectric case, namely,

$$r_p = \frac{-\mathcal{N} \cos \theta + \cos \phi}{\mathcal{N} \cos \theta + \cos \phi} \qquad (TM \ polarization) \qquad (6.81)$$

The derivation is left as an exercise. Knowing the amplitudes of the reflected waves, the amplitudes of the refracted waves can be found from the boundary conditions.

The general behavior of the reflectance as calculated from the above theory is shown graphically in Figure 6.6, in which $R_s = |r_s|^2$

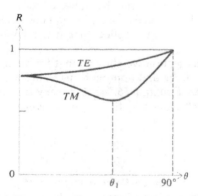

Figure 6.6. Reflectance as a function of angle of incidence for a typical metal.

and $R_p = |r_p|^2$ are plotted as functions of θ for the case of a typical metal. The reflectance for TE polarization increases monotonically from its value at normal incidence, to unity for grazing incidence. On the other hand, for TM polarization the reflectance goes through a shallow minimum for some angle θ_1 whose value depends on the optical constants. This angle is called the *principal angle of incidence* and corresponds to the Brewster angle for dielectrics.

If linearly polarized light, whose polarization is neither pure *TE* nor pure *TM*, is reflected from a metal, then the reflected light will be elliptically polarized, in general. The intensity and the polarization of the reflected light can be calculated from the above theory. This involves knowing the complex index of refraction $\mathcal{N}$. Conversely, it is possible to determine $\mathcal{N}$ by appropriate measurements of the intensity and polarization of the reflected light. The method is known as *ellipsometry*. For more information the reader is referred to Reference [5].

Normal Incidence   In the case of normal incidence, both Equations (6.80) and (6.81) reduce to the same result, namely,

$$r_s = r_p = \frac{1 - \mathcal{N}}{1 + \mathcal{N}} = \frac{1 - n - i\kappa}{1 + n + i\kappa} \tag{6.82}$$

The following expression for the normal reflectance is then obtained:

$$R = \left| \frac{1 - \mathcal{N}}{1 + \mathcal{N}} \right|^2 = \frac{(1 - n)^2 + \kappa^2}{(1 + n)^2 + \kappa^2} \tag{6.83}$$

This reduces to the previously found value for dielectrics (Section 2.7) as κ approaches zero and the index of refraction becomes real. On the other hand, for metals the extinction coefficient κ is large. This results in a high value of the reflectance which approaches unity as κ becomes infinite.

In the previous section we showed that for metals, both n and κ in fact become very large and approach the value $\sqrt{\sigma/2\omega\epsilon_0}$ in the limit of low frequencies, Equation (6.51). It is easy to show from Equation (6.83) that the reflectance in this case is given by the approximate formula

$$R \approx 1 - \frac{2}{n} \approx 1 - \sqrt{\frac{8\omega\epsilon_0}{\sigma}} \tag{6.84}$$

This is known as the *Hagen-Rubens formula*. It has been verified experimentally for a number of metals in the far infrared. The formula predicts that for a given metal the quantity $(1 - R)^2$ is proportional to the frequency or inversely proportional to the wavelength. All of the good conductors − copper, silver, gold, and so on − are excellent reflectors in the near infrared region ($\lambda \approx 1$ to $2\ \mu$) and are even better reflectors in the far infrared ($\lambda > 20\ \mu$), where the reflectance becomes very nearly unity.

6.7 Propagation of Light in Crystals

The distinguishing basic feature of the crystalline state, as far as optical properties are concerned, is the fact that crystals are generally electrically anisotropic. This means that the polarization produced in the crystal by a given electric field is not just a simple scalar constant times the field, but varies in a manner that depends on the direction of the applied field in relation to the crystal lattice. One of the consequences is that the speed of propagation of a light wave in a crystal is a function of the direction of propagation and the polarization of the light.

It turns out that there are generally *two* possible values of the phase velocity for a given direction of propagation. These two values are associated with mutually orthogonal polarizations of the light waves. Crystals are said to be *doubly refracting* or *birefringent*. Actually not all crystals exhibit double refraction. Whether they do or do not depends on their symmetry. Crystals of the cubic class of symmetry, such as sodium chloride, never exhibit double refraction, but are optically isotropic. All crystals, other than cubic crystals, do show double refraction, however.

A model to illustrate the anisotropic polarizability of a crystal is shown in Figure 6.7. A bound electron is pictured here as attached to

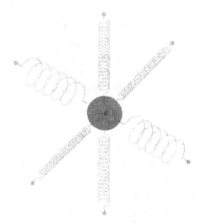

Figure 6.7. Model to show anisotropic binding of an electron in a crystal.

a set of fictitious elastic springs. The springs have different stiffnesses for different directions of the electron's displacement from its equilibrium position within the crystal lattice. Consequently, the displacement of the electron under the action of an external field **E** depends on the direction of the field as well as its magnitude. This is also true of the resulting polarization **P**.

The dependence of **P** on **E** is expressible as a tensor relation in the form

$$\begin{bmatrix} P_x \\ P_y \\ P_z \end{bmatrix} = \epsilon_0 \begin{bmatrix} \chi_{11} & \chi_{12} & \chi_{13} \\ \chi_{21} & \chi_{22} & \chi_{23} \\ \chi_{31} & \chi_{32} & \chi_{33} \end{bmatrix} \begin{bmatrix} E_x \\ E_y \\ E_z \end{bmatrix} \tag{6.85}$$

This is customarily abbreviated as

$$\mathbf{P} = \epsilon_0 \chi \mathbf{E} \tag{6.86}$$

where χ is the susceptibility tensor:

$$\chi = \begin{bmatrix} \chi_{11} & \chi_{12} & \chi_{13} \\ \chi_{21} & \chi_{22} & \chi_{23} \\ \chi_{31} & \chi_{32} & \chi_{33} \end{bmatrix} \tag{6.87}$$

The corresponding displacement vector **D** is given by $\mathbf{D} = \epsilon_0(1 + \chi)\mathbf{E} = \epsilon\mathbf{E}$, where **1** is the unit matrix $\begin{bmatrix} 1 & 0 & 0 \\ 0 & 1 & 0 \\ 0 & 0 & 1 \end{bmatrix}$, and

$$\epsilon = \epsilon_0(1 + \chi) \tag{6.88}$$

which is known as the *dielectric tensor*.

For ordinary nonabsorbing crystals the χ tensor is symmetric so there always exists a set of coordinate axes, called *principal axes,* such that the χ tensor assumes the diagonal form

$$\chi = \begin{bmatrix} \chi_{11} & 0 & 0 \\ 0 & \chi_{22} & 0 \\ 0 & 0 & \chi_{33} \end{bmatrix} \tag{6.89}$$

The three χs are known as the *principal susceptibilities.* Corresponding to these, the quantities $K_{11} = 1 + \chi_{11} \ldots$, and so forth, are called the principal *dielectric constants.*

In view of Equation (6.86), the general wave equation (6.14) can be written in the following form:

$$\nabla \times (\nabla \times \mathbf{E}) + \frac{1}{c^2} \frac{\partial^2 \mathbf{E}}{\partial t^2} = -\frac{1}{c^2} \chi \frac{\partial^2 \mathbf{E}}{\partial t^2} \tag{6.90}$$

It then follows that the crystal can sustain monochromatic plane waves of the usual form $e^{i(\mathbf{k}\cdot\mathbf{r}-\omega t)}$ provided the propagation vector **k** satisfies the equation

$$\mathbf{k} \times (\mathbf{k} \times \mathbf{E}) + \frac{\omega^2}{c^2} \mathbf{E} = -\frac{\omega^2}{c^2} \chi \mathbf{E} \tag{6.91}$$

Written out in terms of components, the above equation is equivalent to the following three equations:

$$\left(-k_y{}^2 - k_z{}^2 + \frac{\omega^2}{c^2}\right) E_x + k_x k_y E_y + k_x k_z E_z = -\frac{\omega^2}{c^2} \chi_{11} E_x$$

$$k_y k_x E_x + \left(-k_x{}^2 - k_z{}^2 + \frac{\omega^2}{c^2}\right) E_y + k_y k_z E_z = -\frac{\omega^2}{c^2} \chi_{22} E_y \tag{6.92}$$

$$k_z k_x E_x + k_z k_y E_y + \left(-k_x{}^2 - k_y{}^2 + \frac{\omega^2}{c^2}\right) E_z = -\frac{\omega^2}{c^2} \chi_{33} E_z$$

In order to interpret the physical meaning of these equations, suppose we have a particular case of a wave propagating in the direction of one of the principal axes, say the x axis. In this case $k_x = k$, $k_y = k_z = 0$, and the three equations reduce to

$$\frac{\omega^2}{c^2} E_x = -\frac{\omega^2}{c^2} \chi_{11} E_x$$

$$\left(-k^2 + \frac{\omega^2}{c^2}\right) E_y = -\frac{\omega^2}{c^2} \chi_{22} E_y \tag{6.93}$$

$$\left(-k^2 + \frac{\omega^2}{c^2}\right) E_z = -\frac{\omega^2}{c^2} \chi_{33} E_z$$

The first equation implies that $E_x = 0$, because neither ω nor χ_{11} is zero. This means that the $\mathbf{E}$ field is transverse to the x axis, which is the direction of propagation. Consider next the second equation. If $E_y \neq 0$, then

$$k = \frac{\omega}{c} \sqrt{1 + \chi_{22}} = \frac{\omega}{c} \sqrt{K_{22}} \tag{6.94}$$

The third equation, likewise, implies that if $E_z \neq 0$, then

$$k = \frac{\omega}{c} \sqrt{1 + \chi_{33}} = \frac{\omega}{c} \sqrt{K_{33}} \tag{6.95}$$

Now ω/k is the phase velocity of the wave. Thus we have two possible phase velocities, namely, $c/\sqrt{K_{22}}$ if the $\mathbf{E}$ vector points in the y direction, and $c/\sqrt{K_{33}}$ if the $\mathbf{E}$ vector is in the z direction.

More generally we can show that for any direction of the propagation vector $\mathbf{k}$, there are two possible values of the magnitude k and hence two possible values of the phase velocity. To do this, let us introduce the three *principal indices of refraction* n_1, n_2, and n_3, defined by

$$n_1 = \sqrt{1 + \chi_{11}} = \sqrt{K_{11}}$$
$$n_2 = \sqrt{1 + \chi_{22}} = \sqrt{K_{22}} \tag{6.96}$$
$$n_3 = \sqrt{1 + \chi_{33}} = \sqrt{K_{33}}$$

Now in Equation (6.92), in order for a nontrivial solution for E_x, E_y, and E_z to exist, the determinant of the coefficients must vanish, namely,

$$\begin{vmatrix} (n_1\omega/c)^2 - k_y^2 - k_z^2 & k_x k_y & k_x k_z \\ k_y k_x & (n_2\omega/c)^2 - k_x^2 - k_z^2 & k_y k_z \\ k_z k_x & k_z k_y & (n_3\omega/c)^2 - k_x^2 - k_y^2 \end{vmatrix} = 0$$

(6.97)

where we have used Equation (6.96). The above equation can be represented by a three-dimensional surface in **k** space. The form of this **k** surface, or wave-vector surface, is shown in Figure 6.8. To see

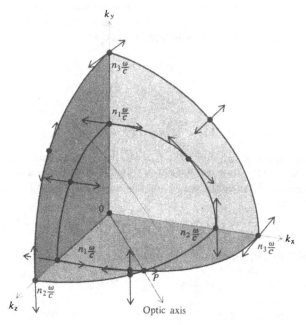

Figure 6.8.   The wave-vector surface.

how the surface is constructed, consider any one of the coordinate planes, say the xy plane. In this plane $k_z = 0$, and the determinant reduces to the product of the two factors

$$\left[\left(\frac{n_3\omega}{c}\right)^2 - k_x^2 - k_y^2\right]\left\{\left[\left(\frac{n_1\omega}{c}\right)^2 - k_y^2\right]\left[\left(\frac{n_2\omega}{c}\right)^2 - k_x^2\right] - k_x^2 k_y^2\right\} = 0$$

(6.98)

Since the product must vanish, either or both of the factors must be equal to zero. Setting the first factor equal to zero gives the equation of a circle

$$k_x{}^2 + k_y{}^2 = \left(\frac{n_3\omega}{c}\right)^2 \tag{6.99}$$

The second factor gives the equation of an ellipse

$$\frac{k_x{}^2}{(n_2\omega/c)^2} + \frac{k_y{}^2}{(n_1\omega/c)^2} = 1 \tag{6.100}$$

Similar equations are obtained for the xz and the yz planes. The intercept of the k surface with each coordinate plane therefore consists of one circle and one ellipse as shown. The complete k surface is double; that is, it consists of an inner sheet and an outer sheet. This implies that for any given direction of the wave vector k, there are two possible values for the wavenumber k. It follows that there are also two values of the phase velocity. Now we just showed that for a wave propagating in the x direction, the two phase velocities correspond to two mutually orthogonal directions of polarization. It turns out that the same is true for any direction of propagation; that is, the two phase velocities always correspond to two mutually orthogonal polarizations [5]. Now, as we know, a light wave of arbitrary polarization can always be resolved into two orthogonally polarized waves. Hence, when unpolarized light, or light of arbitrary polarization propagates through a crystal, it can be considered to consist of two independent waves that are polarized orthogonally with respect to each other and traveling with different phase velocities.

The nature of the k surface is such that the inner and outer sheets touch at a certain point P as shown in Figure 6.8. This point defines a direction for which the two values of k are equal. The direction so defined is called an *optic axis* of the crystal. Thus, when propagating in the direction of an optic axis, the phase velocities of the two orthogonally polarized waves reduce to the same value.

The general case is shown in Figures 6.8 and 6.9(a). Here the three principal indices n_1, n_2, and n_3 are all different. It is easy to see from the intercepts that there are *two* optic axes. In this case the crystal is said to be *biaxial*. In many crystals it happens that two of the principal indices are equal, in which case there is only *one* optic axis and the crystal is called *uniaxial*. The k surface for a uniaxial crystal consists of a sphere and an ellipsoid of revolution, the axis of which is the optic axis of the crystal [Figure 6.9(b) and (c)]. If all three indices are equal, then the k surface degenerates to a single sphere, and the crystal is not doubly refracting at all but is optically isotropic.

In view of the fact that the principal indices are related to the

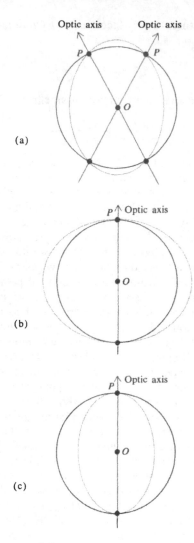

Figure 6.9. Intercepts of the wave-vector surfaces in the xz plane for (a) biaxial crystals; (b) uniaxial positive crystals; (c) uniaxial negative crystals.

components of the χ tensor by Equation (6.96), we can conveniently classify crystals according to the χ tensor as follows:

Isotropic cubic	$\chi = \begin{bmatrix} a & 0 & 0 \\ 0 & a & 0 \\ 0 & 0 & a \end{bmatrix}$	$\chi_{11} = \chi_{22} = \chi_{33} = a$ $n = \sqrt{1+a}$
Uniaxial trigonal tetragonal hexagonal	$\chi = \begin{bmatrix} a & 0 & 0 \\ 0 & a & 0 \\ 0 & 0 & b \end{bmatrix}$	$\chi_{11} = \chi_{22} = a,\ \chi_{33} = b$ $n_O = \sqrt{1+a}$ $n_E = \sqrt{1+b}$
Biaxial triclinic monoclinic orthorhombic	$\chi = \begin{bmatrix} a & 0 & 0 \\ 0 & b & 0 \\ 0 & 0 & c \end{bmatrix}$	$\chi_{11} = a \quad \chi_{22} = b \quad \chi_{33} = c$ $n_1 = \sqrt{1+a}$ $n_2 = \sqrt{1+b}$ $n_3 = \sqrt{1+c}$

In a uniaxial crystal the index of refraction that corresponds to the two equal elements, $\chi_{11} = \chi_{22}$, is called the *ordinary* index n_O, and the other index, corresponding to χ_{33}, is called the *extraordinary* index n_E. If $n_O < n_E$, the crystal is said to be *positive;* whereas if $n_O > n_E$, it is called a *negative* crystal. Table 6.1 lists some examples of crystals with their indices of refraction.

Phase-Velocity Surface[2] Knowing that the wavenumber k is related to the magnitude v of the phase velocity by $k = \omega/v$, we can write the relation vectorially as

$$\mathbf{k} = \mathbf{v}\,\frac{\omega}{v^2} \tag{6.101}$$

In terms of components the above vector equation is equivalent to the three scalar equations

$$k_x = v_x\,\frac{\omega}{v^2} \qquad k_y = v_y\,\frac{\omega}{v^2} \qquad k_z = v_z\,\frac{\omega}{v^2} \tag{6.102}$$

Let us substitute the above values into the equation of the **k** surface in Equation (6.97). The result is

$$\begin{vmatrix} n_1^2 v^4/c^2 - v_y^2 - v_z^2 & v_x v_y & v_x v_z \\ v_y v_x & n_2^2 v^4/c^2 - v_x^2 - v_z^2 & v_y v_z \\ v_z v_x & v_z v_y & n_3^2 v^4/c^2 - v_x^2 - v_y^2 \end{vmatrix} = 0 \tag{6.103}$$

[2] In this section we depart from the notation used in the other parts of the book, namely, we use this letter v for phase velocity, rather than u, which will be used here for ray velocity.

Table 6.1. SOME COMMON CRYSTALS

─────────OPTICALLY ISOTROPIC (CUBIC) CRYSTALS─────────	
	n
Sodium chloride	1.544
Diamond	2.417
Fluorite	1.392

──────────UNIAXIAL POSITIVE CRYSTALS──────────		
	n_O	n_E
Ice	1.309	1.310
Quartz	1.544	1.553
Zircon	1.923	1.968
Rutile	2.616	2.903

────────UNIAXIAL NEGATIVE CRYSTALS────────		
	n_O	n_E
Beryl	1.598	1.590
Sodium nitrate	1.587	1.336
Calcite	1.658	1.486
Tourmaline	1.669	1.638

────────BIAXIAL CRYSTALS────────			
	n_1	n_2	n_3
Gypsum	1.520	1.523	1.530
Feldspar	1.522	1.526	1.530
Mica	1.552	1.582	1.588
Topaz	1.619	1.620	1.627

after cancellation of ω^2 and division by v^4. This equation defines a three-dimensional surface that can be considered as the *reciprocal surface* to the **k** surface. It is called the *phase-velocity surface*. It is a double-sheeted surface and gives directly the two possible values of the phase velocity for a given direction of a plane wave propagating in the crystal. The general form of the phase-velocity surface is shown in Figure 6.10. The intercepts with the coordinate planes consists of circles and *fourth-degree ovals*. Thus for the xy plane the two equations for the intercepts are

$$v^2 = v_x{}^2 + v_y{}^2 = \frac{c^2}{n_3{}^2} \tag{6.104}$$

$$\frac{v_x{}^2}{n_2{}^2} + \frac{v_y{}^2}{n_1{}^2} = \frac{v^4}{c^2} \tag{6.105}$$

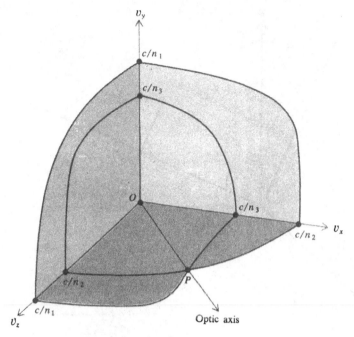

Figure 6.10. The phase-velocity surface.

Similar relations hold for the other coordinate planes.

The Poynting Vector and the Ray Velocity   Although the propagation vector **k** defines the direction of the planes of constant phase for light waves in a crystal, the actual direction of the energy flow **E** × **H** is not in the same direction, generally, as that of **k**. This stems from the fact that in anisotropic media **E** and **k** are not, in general, mutually perpendicular, as can be seen by inspection of Equation (6.91). On the other hand the magnetic field **H** is perpendicular to both **E** and **k** because of the relation **k** × **E** = $\mu_0 \omega$**H**, which comes from the first Maxwell equation. The situation is shown graphically in Figure 6.11. The three vectors **E**, **k**, and **S** = **E** × **H** are all perpendicular to **H** and, further, **E** is perpendicular to **S**.

Consider a narrow beam or ray of light in a crystal. The planes of constant phase are perpendicular to **k**, but they move along the direction of the ray **S**. Thus the planes of constant phase are inclined to their direction of motion as shown in the figure. Let θ denote the angle between **k** and **S**. Then the surfaces of constant phase move with a velocity u — called the *ray velocity* — along the ray direction.

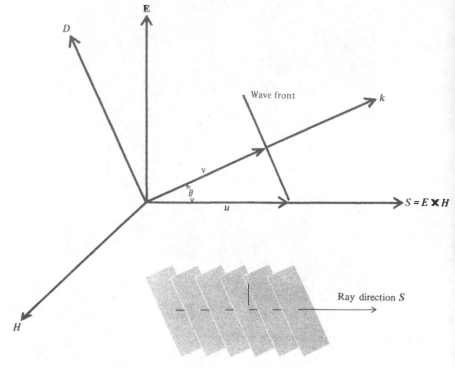

Figure 6.11. Relationships among the electric and magnetic fields, the Poynting vector, the wave vector, the ray-velocity vector, and the phase-velocity vector for plane waves in a crystal.

The magnitude of u is given by

$$u = \frac{v}{\cos \theta} \tag{6.106}$$

where v is the phase velocity (in the direction of k). Evidently the ray velocity is greater than the phase velocity except, of course, when $\theta = 0$. In the latter case the phase and ray velocities are equal. This occurs when the direction of propagation is along one of the principal axes of the crystal. In this case $\mathbf{S}$ and $\mathbf{k}$ also have the same direction.

The Ray-Velocity Surface   This surface gives the magnitude of the ray velocity for any given direction of the ray. To find the equation of the ray-velocity surface, it will be convenient to express the wave Equation (6.91) in terms of the displacement vector $\mathbf{D} = \epsilon_0(1 + \chi)\mathbf{E}$. We have, then, for plane harmonic waves

$$\mathbf{k} \times (\mathbf{k} \times \mathbf{E}) = -\frac{\omega^2}{c^2\epsilon_0} \mathbf{D}$$

This shows that **D** is perpendicular to the wave vector **k**. Expanding the triple product gives

$$\mathbf{k}(\mathbf{k} \cdot \mathbf{E}) - k^2 \mathbf{E} = -\frac{\omega^2}{c^2 \epsilon_0} \mathbf{D}$$

Next, take the dot product with **D**. Then, since $\mathbf{k} \cdot \mathbf{D} = 0$, we find

$$k^2 \mathbf{E} \cdot \mathbf{D} = \frac{\omega^2}{c^2 \epsilon_0} \mathbf{D} \cdot \mathbf{D}$$

or, in view of the fact that $v = \omega/k$,

$$\mathbf{E} \cdot \mathbf{D} = ED \cos \theta = \frac{v^2}{c^2 \epsilon_0} D^2$$

Now if the coordinate axes are principal axes of the crystal, the components of **E** are related to those of **D** by

$$\epsilon_0 E_x = \frac{D_x}{\epsilon_{11}} = \frac{D_x}{n_1^2}$$

and similarly for the y and z components. Consequently, the equation for $\mathbf{E} \cdot \mathbf{D}$ above is equivalent to the following three scalar equations:

$$D_x \left(\frac{c^2}{n_1^2} - u_y^2 - u_z^2 \right) + D_y u_x u_y + D_z u_x u_z = 0$$

$$D_x u_y u_x + D_y \left(\frac{c^2}{n_2^2} - u_x^2 - u_z^2 \right) + D_z u_y u_z = 0$$

$$D_x u_z u_x + D_y u_y u_z + D_z \left(\frac{c^2}{n_3^2} - u_x^2 - u_y^2 \right) = 0$$

The determinant of the coefficients must vanish in order that a nontrivial solution exists. This gives the equation of the ray-velocity surface,

$$\begin{vmatrix} c^2/n_1^2 - u_y^2 - u_z^2 & u_x u_y & u_x u_z \\ u_y u_x & c^2/n_2^2 - u_x^2 - u_z^2 & u_y u_z \\ u_z u_x & u_z u_y & c^2/n_3^2 - u_x^2 - u_y^2 \end{vmatrix} = 0 \qquad (6.107)$$

In particular, the equations of the intercepts in the xy plane are obtained by setting $u_z = 0$. The result gives a circle

$$u_x^2 + u_y^2 = \frac{c^2}{n_3^2} \qquad (6.108)$$

and an ellipse

$$n_2^2 u_x^2 + n_1^2 u_y^2 = c^2 \qquad (6.109)$$

Corresponding equations can be obtained for the other coordinate planes by cyclic permutation, and in each case the intercepts consist of an ellipse and a circle. It is easily verified that the intercepts of the

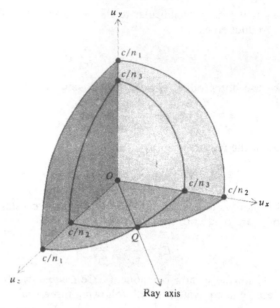

Figure 6.12. The ray-velocity surface.

ray-velocity surface along the coordinate axes are the same as those of the phase-velocity surface. Figure 6.12 shows the form of the ray-velocity surface. As with the phase-velocity surface, the ray-velocity surface consists of two sheets, an inner one and an outer one, corresponding to the two possible values of u for a given ray direction. The two sheets touch at a point Q that defines a direction for which the two ray velocities are equal. This direction is called the *ray axis* of the crystal.

In biaxial crystals there are two ray axes, and these ray axes are distinct from the optic axes of the crystal. On the other hand, in the case of a uniaxial crystal the two sheets of the ray-velocity surface consist of a sphere and an ellipsoid of revolution (spheroid). The two surfaces are tangent at the ends of a particular diameter of the sphere. This diameter defines the ray axis that, for uniaxial crystals, also coincides with the optic axis of the crystal.

6.8 Double Refraction at a Boundary

Consider a plane wave incident on the surface of a crystal. Denote the propagation vector of the incident wave by $\mathbf{k}_0$ and that of the refracted wave by $\mathbf{k}$, and call θ and ϕ the angles of incidence and refraction. According to the arguments in Section 2.6, where we treated refraction at a dielectric boundary, the law of refraction was seen to be contained in the equation

$$\mathbf{k_0} \cdot \mathbf{r} = \mathbf{k} \cdot \mathbf{r} \qquad \text{(at boundary)} \qquad (6.110)$$

This same relation is also true for refraction at a boundary of a crystal since it merely expresses the fact that some unspecified boundary condition can exist at all. The equation implies that the projections of the propagation vectors along the boundary plane must be equal for both the incident and refracted waves. Now we know that for a given direction of propagation in the crystal, there are two possible propagation vectors. Owing to the double nature of the $\mathbf{k}$ surface, it is also true that for a prescribed value of the *projection* of the propagation vector in any given direction there are again two possible propagation vectors. This results in double refraction of a wave incident on the surface of a crystal as shown in Figure 6.13. From Equation (6.110),

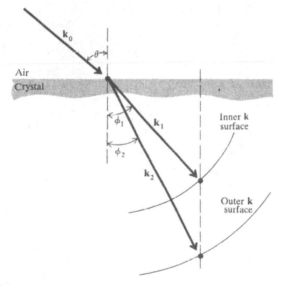

Figure 6.13. Wave vectors for double refraction at the boundary of a crystal.

we can write

$$k_0 \sin \theta = k_1 \sin \phi_1 \qquad k_0 \sin \theta = k_2 \sin \phi_2 \qquad (6.111)$$

for the two refracted waves.

At first sight it may appear that the above equations constitute a statement of Snell's law for double refraction. However, this is not the case. The trouble is that k_1 and k_2 are not constant in general, rather, they vary with the directions of the vectors k_1 and k_2. This means that the ratio $\sin \theta / \sin \phi$ is not always constant as it is in the case of refraction at the boundary of an isotropic medium. The problem of determining ϕ, given the value of θ, is thus not a simple one. One way is to solve for ϕ graphically as suggested by Figure 6.13.

In the case of uniaxial crystals, as we have seen, one of the parts of the k surface is a sphere. The corresponding wavenumber k is constant for all directions of the wave in the crystal, and Snell's law is obeyed. This wave is known as the *ordinary wave*, and we have

$$\frac{\sin \theta}{\sin \phi} = n_O \qquad (6.112)$$

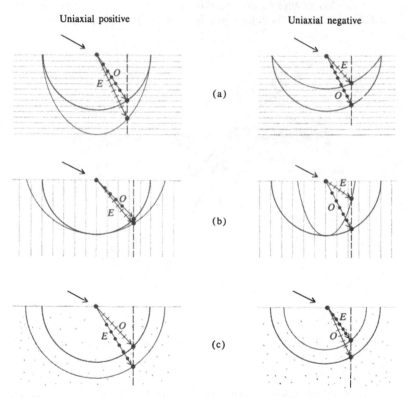

Figure 6.14. Wave vectors for double refraction in uniaxial crystals. (a) The optic axis parallel to the boundary and parallel to the plane of incidence; (b) the optic axis perpendicular to the boundary and parallel to the plane of incidence; (c) the optic axis parallel to the boundary and perpendicular to the plane of incidence.

where n_O is the ordinary index of refraction. The **k** surface for the other wave, however, is a spheroid, and Snell's law is not valid. This wave is called the *extraordinary wave*. Since the extraordinary index n_E is greater than n_O for positive uniaxial crystals, and less than n_O for negative crystals, we conclude that $\phi_E \le \phi_o$ for positive crystals and $\phi_E \ge \phi_o$ for negative crystals. Some examples of double refraction are illustrated in Figure 6.14. In all cases the polarizations of the two waves are mutually orthogonal. The directions of the wave vectors are indicated in the figure. The *ray direction*, corresponding to any given wave vector **k**, is found by constructing the normal to the **k** surface at the end point of the **k** vector. A proof is given in Reference [5].

Polarizing Prisms   Let a wave be incident on a plane boundary from the *inside* of a uniaxial crystal. Consider the special case in which the optic axis is perpendicular to the plane of incidence as shown in Figure 6.14(c). Then the cross section of the **k** surface consists of two circles, as shown, and therefore Snell's law holds for both the ordinary wave and the extraordinary wave. For simplicity let the external medium be air ($n = 1$). Then we can write

$$n_O \sin \phi_O = \sin \theta \qquad (6.113)$$

$$n_E \sin \phi_E = \sin \theta \qquad (6.114)$$

where θ is the internal angle of incidence, and ϕ_O and ϕ_E are the angles of refraction of the ordinary wave and the extraordinary wave, respectively. The **E** vector of the ordinary wave is perpendicular to the direction of the optic axis, and the **E** vector of the extraordinary wave is parallel to the optic axis.

Suppose now that we have a negative uniaxial crystal, such as calcite, and that the internal angle of incidence θ is such that

$$n_E < \frac{1}{\sin \theta} < n_0 \qquad (6.115)$$

In this case we have total internal reflection for the ordinary wave but *not* for the extraordinary wave. The refracted wave is thus completely polarized as shown in Figure 6.15(a). This is the basic principle for producing polarized light by means of double refraction.

One of the most commonly used polarizing prisms is the Glan prism shown in Figure 6.15(b). It consists of two identical prisms of calcite cut so that the optic axes are parallel to the corner edges, and mounted so that the long faces are parallel as shown. The space between the two prisms may be air or any suitable transparent material. If an air gap is used, the apex angle must be about 38.5 degrees.

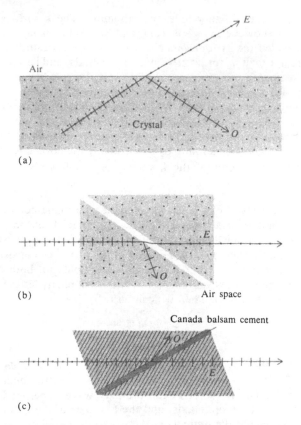

(a)

(b)   Air space

Canada balsam cement

(c)

Figure 6.15. (a) Separation of the extraordinary and ordinary rays at the boundary of a crystal in the case of internal refraction; (b) construction of the Glan polarizing prism; (c) the Nicol prism.

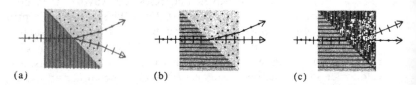

(a)   (b)   (c)

Figure 6.16. Three types of prisms for separating unpolarized light into two divergent orthogonally polarized beams. (a) The Wollaston prism; (b) the Rochon prism; (c) the Senarmont prism. All prisms shown are made with uniaxial positive material (quartz).

An older type of polarizing prism is the Nicol prism. It is made in the form of a rhomb having approximately the same shape as a natural crystal of calcite [Figure 6.15(c)]. The Nicol prism is inferior to the Glan prism in most respects and is largely of historical interest.

Another type of polarizing device makes use of double refraction to separate an incident beam of light into two diverging beams having mutually orthogonal directions of polarization. Three ways of doing this are illustrated in Figure 6.16. The figures are self-explanatory.

6.9   Optical Activity

Certain substances are found to possess the ability to rotate the plane of polarization of light passing through them. This phenomenon is known as *optical activity*. When a beam of linearly polarized light is passed through an optically active medium (Figure 6.17), the light

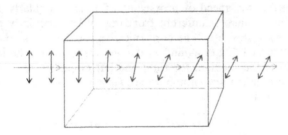

Figure 6.17. Rotation of the plane of polarization by an optically active medium. The case for levorotatory is shown.

emerges with its plane of polarization turned through an angle that is proportional to the length of the path of the light through the medium. The amount of rotation per unit length of travel is called the *specific rotatory power*. If the sense of rotation of the plane of polarization is to the right, as a right-handed screw pointing in the direction of propagation, the substance is called *dextrorotatory* or right handed. If the rotation is to the left, the substance is called *levorotatory* or left handed. Sodium chlorate, cinnabar, and certain kinds of sugar are examples of optically active substances. Fused quartz is optically isotropic, but crystallized quartz is optically active as well as double refracting.

Quartz occurs in two crystalline forms, right handed and left handed. These crystals are found to be dextrorotatory and levorotatory, respectively. The specific rotatory power of either of the two kinds of quartz for light propagating along the direction of the optic axis is tabulated for different wavelengths in Table 6.2. It is seen that

Table 6.2. OPTICAL ACTIVITY OF QUARTZ

Wavelength (Å)	Specific Rotatory Power (degrees/mm)
4000	49
4500	37
5000	31
5500	26
6000	22
6500	17

the amount of optical activity of quartz varies with wavelength. This variation with wavelength is called *rotatory dispersion.*

Optical activity can be explained on the basis of the simple assumption that the speed of propagation for right circularly polarized light in the medium is different from that of left circularly polarized light. To show this, it will be convenient to use the Jones vector notation of Section 2.5. Let n_R and n_L denote, respectively, the indices of refraction of the medium for right and left circularly polarized light. The corresponding wavenumbers are $k_R = n_R\omega/c$ and $k_L = n_L\omega/c$, and the expressions

$$\begin{bmatrix} 1 \\ -i \end{bmatrix} e^{i(k_R z - \omega t)} \qquad \begin{bmatrix} 1 \\ i \end{bmatrix} e^{i(k_L z - \omega t)}$$

represent the two kinds of wave in the medium.

Now suppose that a beam of linearly polarized light travels a distance l through the medium. Let the initial polarization be in the horizontal direction. The initial Jones vector, separated into right and left circular components, is

$$\begin{bmatrix} 1 \\ 0 \end{bmatrix} = \tfrac{1}{2} \begin{bmatrix} 1 \\ -i \end{bmatrix} + \tfrac{1}{2} \begin{bmatrix} 1 \\ i \end{bmatrix}$$

The complex amplitude of the light wave, after traveling a distance l through the medium, is

$$\tfrac{1}{2} \begin{bmatrix} 1 \\ -i \end{bmatrix} e^{ik_R l} + \tfrac{1}{2} \begin{bmatrix} 1 \\ i \end{bmatrix} e^{ik_L l}$$

$$= \tfrac{1}{2}\, e^{i(k_R + k_L)l/2} \left\{ \begin{bmatrix} 1 \\ -i \end{bmatrix} e^{i(k_R - k_L)l/2} + \begin{bmatrix} 1 \\ i \end{bmatrix} e^{-i(k_R - k_L)l/2} \right\} \tag{6.116}$$

Upon introducing the quantities ψ and θ where

$$\psi = \tfrac{1}{2}(k_R + k_L)l \tag{6.117}$$

$$\theta = \tfrac{1}{2}(k_R - k_L)l \tag{6.118}$$

we can express the complex amplitude as

$$e^{i\psi} \left\{ \tfrac{1}{2} \begin{bmatrix} 1 \\ -i \end{bmatrix} e^{i\theta} + \tfrac{1}{2} \begin{bmatrix} 1 \\ i \end{bmatrix} e^{-i\theta} \right\}$$

$$= e^{i\psi} \begin{bmatrix} \tfrac{1}{2}(e^{i\theta} + e^{-i\theta}) \\ \tfrac{1}{2}i(e^{i\theta} - e^{-i\theta}) \end{bmatrix} \qquad (6.119)$$

$$= e^{i\psi} \begin{bmatrix} \cos\theta \\ \sin\theta \end{bmatrix}$$

This represents a linearly polarized wave in which the direction of polarization is turned through an angle θ with respect to the original direction of polarization. From Equation (6.118) we have

$$\theta = (n_R - n_L) \frac{\omega l}{2c} = (n_R - n_L) \frac{\pi l}{\lambda} \qquad (6.120)$$

where λ is the wavelength in vacuum. It follows that the specific rotatory power δ, as a function of wavelength, is given by

$$\delta = (n_R - n_L) \frac{\pi}{\lambda} \qquad (6.121)$$

The indices n_R and n_L are also, of course, functions of wavelength.

As a numerical example, the indices for propagation along the optic axis in quartz are as follows:

λ	n_R	n_L	$n_R - n_L$
3960 Å	1.55810	1.55821	0.00011
5890 Å	1.54420	1.54427	0.00007
7600 Å	1.53914	1.53920	0.00006

These are for right-handed quartz. For left-handed quartz the values are just reversed.

A method of separating unpolarized light into two beams of oppositely rotating circularly polarized light was devised by Fresnel. Two prisms made of right-handed and left-handed quartz are arranged as shown in Figure 6.18. The relative index of refraction at the

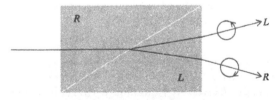

Figure 6.18. The Fresnel prism for separating unpolarized light into two divergent beams of opposite circular polarizations.

diagonal boundary is greater than unity for right polarized light and less than unity for left polarized light. Hence the beam is separated into two beams at the boundary as shown. The light emerges from the prism in two divergent beams. This prism can also be used to determine the sense of rotation of circularly polarized light.

Susceptibility Tensor of an Optically Active Medium   It is a simple matter to show that if the susceptibility tensor has conjugate imaginary off-diagonal elements, namely,

$$\chi = \begin{bmatrix} \chi_{11} & i\chi_{12} & 0 \\ -i\chi_{12} & \chi_{11} & 0 \\ 0 & 0 & \chi_{33} \end{bmatrix} \tag{6.122}$$

where χ_{12} is real, then the medium is optically active. To prove this we write the components of the wave equation (6.91) for the above susceptibility tensor. For simplicity we consider the case of a wave propagating in the z direction. Then we have

$$-k^2 E_x + \frac{\omega^2}{c^2} E_x = -\frac{\omega^2}{c^2} (\chi_{11} E_x + i\chi_{12} E_y) \tag{6.123}$$

$$-k^2 E_y + \frac{\omega^2}{c^2} E_y = -\frac{\omega^2}{c^2} (-i\chi_{12} E_x + \chi_{11} E_y) \tag{6.124}$$

$$\frac{\omega^2}{c^2} E_z = -\frac{\omega^2}{c^2} \chi_{33} E_z \tag{6.125}$$

The last equation merely gives $E_z = 0$, so the wave is transverse. The determinant of the coefficients of the first two equations must vanish for a nontrivial solution, namely,

$$\begin{vmatrix} -k^2 + (\omega^2/c^2)(1 + \chi_{11}) & i(\omega^2/c^2)\chi_{12} \\ -i(\omega^2/c^2)\chi_{12} & -k^2 + (\omega^2/c^2)(1 + \chi_{11}) \end{vmatrix} = 0 \tag{6.126}$$

Solving for k, we find

$$k = \frac{\omega}{c} \sqrt{1 + \chi_{11} \pm \chi_{12}} \tag{6.127}$$

Now if we substitute the above expression for k back into either of Equations (6.123) or (6.124), we obtain

$$E_x = \pm i E_y \tag{6.128}$$

where the upper sign corresponds to the upper sign in Equation (6.127) and similarly for the lower sign. The above result means that the two values of k given by Equation (6.127) correspond to right and left circularly polarized light. The indices of refraction are, accordingly,

$$n_R = \sqrt{1 + \chi_{11} + \chi_{12}} \qquad (6.129)$$

$$n_L = \sqrt{1 + \chi_{11} - \chi_{12}} \qquad (6.130)$$

for right and left circularly polarized light, respectively. It follows that the difference between n_R and n_L is given approximately by

$$n_R - n_L \approx \frac{\chi_{12}}{\sqrt{1 + \chi_{11}}} = \frac{\chi_{12}}{n_0} \qquad (6.131)$$

where n_0 is the *ordinary* index of refraction. The specific rotatory power from Equation (6.121) is then

$$\delta = \frac{\chi_{12}\pi}{n_0\lambda} \qquad (6.132)$$

Our result shows that the specific rotatory power is directly proportional to the imaginary component χ_{12} of the susceptibility tensor.

The k Surface for Quartz Crystalline quartz is optically active as well as doubly refracting. Thus the susceptibility tensor for quartz is of the form given by Equation (6.122) rather than that of a simple uniaxial crystal. The correct equation of the **k** surface for quartz is thus of the form

$$\begin{vmatrix} (n_1\omega/c)^2 - k_y^2 - k_z^2 & k_x k_y + i\chi_{12}(\omega/c)^2 & k_x k_z \\ k_y k_x - i\chi_{12}(\omega/c)^2 & (n_1/c)^2 - k_z^2 - k_x^2 & k_y k_z \\ k_z k_x & k_z k_y & (n_3\omega/c)^2 - k_x^2 - k_y^2 \end{vmatrix} = 0$$

$$(6.133)$$

A plot of the surface is shown in Figure 6.19. The two sheets of the k surface no longer refer to orthogonal linear polarizations, but rather to orthogonal *elliptical* polarizations. The type of polarization is indicated on the figure for various directions of propagation. Along the direction of the optic axis, the inner and outer surfaces do not touch (as they do in the case of an ordinary uniaxial crystal) but are separated by a certain amount. The separation depends on the value of χ_{12} and is therefore a measure of the optical rotatory power.

6.10 Faraday Rotation in Solids

If an isotropic dielectric is placed in a magnetic field and a beam of linearly polarized light is sent through the dielectric in the direction of the field, a rotation of the plane of polarization of the emerging light is found to occur. In other words the presence of the field causes the dielectric to become optically active. This phenomenon was discovered in 1845 by Michael Faraday. The amount of rotation θ of the plane of polarization of the light is proportional to the magnetic induc-

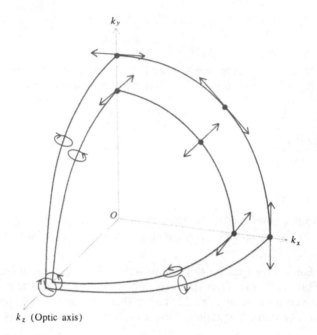

Figure 6.19. The wave-vector surface of quartz.

tion B and to the length l of travel in the medium. Thus we may write

$$\theta = VBl \qquad (6.134)$$

where V is a constant of proportionality. This constant is called the *Verdet constant*. Some examples are tabulated in Table 6.3. The figures given are for yellow light, 5890 Å.

Table 6.3. VALUES OF THE VERDET CONSTANT FOR SOME SELECTED SUBSTANCES

Substance	V (Minutes of Angle/per Oe/cm)
Fluorite	0.0009
Diamond	0.012
Glass	
Crown	0.015–0.025
Flint	0.030–0.050
Sodium chloride	0.036

In order to explain the Faraday effect, we must consider the equation of motion of the bound electrons in the presence of the static magnetic field **B** and the oscillating electric field **E** of the light wave. The differential equation of motion is

$$m \frac{d^2\mathbf{r}}{dt^2} + K\mathbf{r} = -e\mathbf{E} - e \left(\frac{d\mathbf{r}}{dt}\right) \times \mathbf{B} \qquad (6.135)$$

where, as in the treatment of the theory of dispersion in dielectric media (Section 6.4), **r** is the displacement of the electron from its equilibrium position and K is the elastic-force constant. For reasons of simplicity we neglect the force due to the magnetic field of the optical wave as well as the damping effect. These small effects are not particularly germane to the understanding of the basic theory of the Faraday effect.

We assume that the optical field **E** has the usual harmonic time dependence $e^{-i\omega t}$. The particular solution that we are interested in is the steady state condition for which the displacement **r** has the same harmonic time dependence as the light wave. Hence we can write

$$-m\omega^2 \mathbf{r} + K\mathbf{r} = -e\mathbf{E} + i\omega e\mathbf{r} \times \mathbf{B} \qquad (6.136)$$

But the polarization **P** of the medium is just a constant times **r**, namely, $-N e \mathbf{r}$, hence the above equation implies that

$$(-m\omega^2 + K)\mathbf{P} = N e^2 \mathbf{E} + i\omega e \mathbf{P} \times \mathbf{B} \qquad (6.137)$$

Now this equation can be solved for **P** by writing the equation in component form and solving for the components of **P**. The result is expressible in the normal way:

$$\mathbf{P} = \epsilon_0 \chi \mathbf{E} \qquad (6.138)$$

where χ is the "effective" susceptibility tensor. Its form is precisely that of an optically active medium, namely,

$$\chi = \begin{bmatrix} \chi_{11} & +i\chi_{12} & 0 \\ -i\chi_{12} & \chi_{11} & 0 \\ 0 & 0 & \chi_{33} \end{bmatrix} \qquad (6.139)$$

where

$$\chi_{11} = \frac{N e^2}{m\epsilon_0} \left[\frac{\omega_0{}^2 - \omega^2}{(\omega_0{}^2 - \omega^2)^2 - \omega^2 \omega_c{}^2} \right] \qquad (6.140)$$

$$\chi_{33} = \frac{N e^2}{m\epsilon_0} \left[\frac{1}{\omega_0{}^2 - \omega^2} \right] \qquad (6.141)$$

$$\chi_{12} = \frac{N e^2}{m\epsilon_0} \left[\frac{\omega \omega_c}{(\omega_0{}^2 - \omega^2)^2 - \omega^2 \omega_c{}^2} \right] \qquad (6.142)$$

In deriving the above result it has been assumed that the magnetic field **B** is in the z direction. The following abbreviations are used:

$$\omega_0 = \sqrt{\frac{K}{m}} \qquad \text{(resonance frequency)} \qquad (6.143)$$

$$\omega_c = \frac{eB}{m} \qquad \text{(cyclotron frequency)} \qquad (6.144)$$

Finally, referring to Equation (6.132), we see that the specific rotatory power induced by a magnetic field is given by the approximate equation

$$\delta \approx \frac{\pi N e^2}{\lambda m \epsilon_0}\left[\frac{\omega \omega_c}{(\omega_0{}^2 - \omega^2)^2}\right] = \frac{\pi N e^3}{\lambda m^2 \epsilon_0}\left[\frac{\omega B}{(\omega_0{}^2 - \omega^2)^2}\right] \qquad (6.145)$$

in which it is assumed that $\omega\omega_c \ll \left|\omega_0{}^2 - \omega^2\right|$.

6.11 Other Magneto-optic and Electro-optic Effects

According to the theory developed in the previous section, a substance becomes doubly refracting as well as optically active in the presence of a static magnetic field. This is because of the fact that χ_{11} and χ_{33} are different. However, this double refraction is very small except when the frequency of the light is near the resonance frequency. Magnetically induced double refraction is observed in atomic vapors at optical frequencies close to the resonance frequencies of the atoms comprising the vapor. This phenomenon is called the *Voigt effect.*

Kerr Electro-optic Effect When an optically isotropic substance is placed in a strong electric field, it becomes doubly refracting. The effect was discovered in 1875 by J. Kerr and is called the *Kerr electro-optic effect.* It is observed in both solids (glass) and liquids. The Kerr electro-optic effect is attributed to the alignment of the molecules in the presence of the electric field. The substance then behaves op-

Table 6.4. VALUES OF THE KERR CONSTANT

Substance	K (cm/V^2)
Benzene	0.7×10^{-12}
Carbon disulfide	3.5×10^{-12}
Nitrotoluene	2.0×10^{-10}
Nitrobenzene	4.4×10^{-10}

tically as if it were a uniaxial crystal in which the electric field defines the optic axis. The magnitude of the effect is found to be proportional to the *square of the electric field strength*. The Kerr constant K is defined by the equation

$$n_\| - n_\perp = KE^2\lambda_0 \tag{6.146}$$

where $n_\|$ is the index of refraction in the direction of the applied field **E**, and $n_\perp$ is the index at right angles to **E**. The vacuum wavelength is λ_0. Table 6.4 lists the Kerr constants of several liquids.

The Kerr electro-optic effect is utilized to produce a high-speed light modulator known as a "Kerr cell." This device, shown in Figure 6.20, consists of two parallel conductors immersed in a suitable

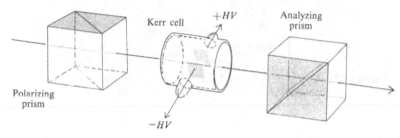

Figure 6.20.   Arrangement for using a Kerr cell light modulator. (Note: The Kerr cell is normally oriented so that the electric vector of the incoming light wave is inclined at 45° with the electric field inside the Kerr cell.)

liquid. (Nitrobenzene is generally used because of its high Kerr constant.) If the polarizer and analyzer are crossed and are oriented at ±45 degrees with respect to the electric axis of the Kerr cell, then no light is transmitted except when the electric field is turned on. The transmission as a function of applied voltage is shown in Figure 6.22.

The Cotton-Mouton Effect   The *Cotton-Mouton effect* is the magnetic analogue of the Kerr electro-optic effect. It is observed in liquids and is attributed to the "lining up" of the molecules by the magnetic field. Like the Kerr effect, the Cotton-Mouton effect is also found to be proportional to the square of the impressed field.

The Pockels Effect   When certain kinds of birefringent crystals are placed in an electric field, their indices of refraction are altered by the presence of the field. This effect is known as the *Pockels effect*. It is found to be directly proportional to an applied field strength. The effect is utilized to produce light shutters and modulators. Pockels cells are commonly made with ADP (ammonium dihydrogen phosphate) or KDP (potassium dihydrogen phosphate). The crystal is placed between electrodes arranged so that light passes in the same direction as the electric field (Figure 6.21). The voltage versus transmission curve is shown in Figure 6.22.

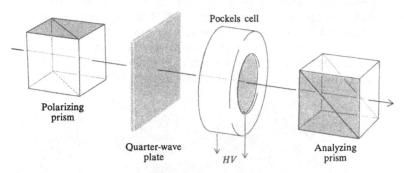

Figure 6.21. Setup for using a Pockels cell light modulator. The quarter-wave plate is used to provide "optical bias."

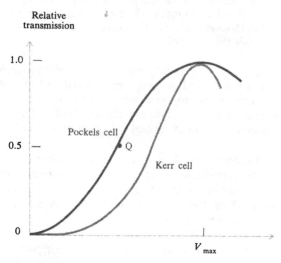

Figure 6.22. Transmission curves of the Kerr cell and Pockels cell. The operating point for the Pockels cell would be at the point marked Q when a quarter-wave bias plate is used, as shown in Figure 6.21.

6.12   Nonlinear Optics

When a light wave propagates through an optical medium, the oscillating electromagnetic field exerts a polarizing force on all of the electrons comprising the medium. Since the inner electrons of the atoms are tightly bound to the nuclei, the major polarizing effect is exerted on the outer or valence electrons. With ordinary light sources the radiation fields are much smaller than the fields that bind the electrons to the atoms. Hence the radiation acts as a small perturbation. This produces a polarization that is proportional to the electric field of the light wave. However, if the radiation field is comparable with the atomic fields ($\sim 10^8$ V/cm), then the relation between the polarization and the radiation field is no longer a linear one (Figure 6.23).

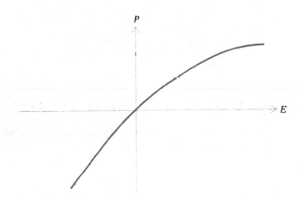

Figure 6.23.   Curve showing polarization versus electric field for a nonlinear dielectric.

The requisite light fields needed to exhibit this nonlinearity are obtainable with laser sources. Nonlinear optical effects that have been observed include optical harmonic generation, production of combination frequencies, optical rectification, and many others [3], [42].

In an isotropic medium the general relation between the polarization **P** and the electric field **E** is expressible as a simple series expansion involving only the magnitudes since the direction of the polarization coincides with that of the field, namely,

$$P = \epsilon_0(\chi E + \chi^{(2)}E^2 + \chi^{(3)}E^3 + \cdots) \qquad (6.147)$$

In this expansion χ is the normal or linear susceptibility. It is generally much larger than the nonlinear coefficients $\chi^{(2)}$, $\chi^{(3)}$, and so forth. If the applied field has the form $E_0 e^{-i\omega t}$, then the induced polarization is

$$P = \epsilon_0(\chi E_0 e^{-i\omega t} + \chi^{(2)}E_0{}^2 e^{-i2\omega t} + \chi^{(3)}E_0{}^3 e^{-i3\omega t} \cdots) \qquad (6.148)$$

The part of the polarization associated with the second and higher terms gives rise to the generation of optical harmonics (Figure 6.24).

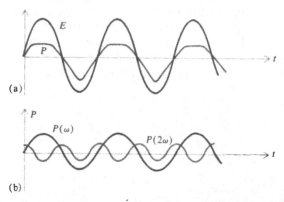

Figure 6.24. (a) Graphs of electric field and polarization as functions of time for the nonlinear case; (b) resolution of the polarization into the fundamental and the second harmonic. (There will also be a dc term that is not shown.)

These usually decrease rapidly in intensity with increasing order. If the general relation between **P** and **E** is such that a reversal of the direction of **E** merely results in the reversal of the direction of **P**, that is, if $P(E)$ is an odd function, then the even terms are all zero and there are no even harmonics. This is, in fact, the case with isotropic media.

In the case of crystalline media, **P** and **E** are not necessarily parallel. The polarization must then be expressed as an expansion of the type

$$\mathbf{P} = \epsilon_0(\chi\mathbf{E} + \chi^{(2)}\mathbf{EE} + \chi^{(3)}\mathbf{EEE} + \cdots) \qquad (6.149)$$

where χ is the ordinary susceptibility tensor. The coefficients $\chi^{(2)}$, $\chi^{(3)}$, and so forth, are higher-order tensors. The expansion is often written as the sum of two terms

$$\mathbf{P} = \mathbf{P}^{\mathrm{L}} + \mathbf{P}^{\mathrm{NL}} \qquad (6.150)$$

where the linear polarization is

$$\mathbf{P}^{\mathrm{L}} = \epsilon_0\chi\mathbf{E} \qquad (6.151)$$

The remainder is the nonlinear polarization and is given by

$$\mathbf{P}^{\text{NL}} = \epsilon_0 \chi^{(2)} \mathbf{EE} + \epsilon_0 \chi^{(3)} \mathbf{EEE} + \cdots \qquad (6.152)$$

If the impressed field $\mathbf{E}$ is a light wave of angular frequency ω, then the second harmonic polarization $P(2\omega)$ arises from the term $\epsilon_0 \chi^{(2)} \mathbf{EE}$. Its components can be written

$$P_i(2\omega) = \sum_j \sum_k \chi_{ijk}^{(2)} E_j E_k \qquad (6.153)$$

The amount of second harmonic light that is produced depends critically on the form of the $\chi^{(2)}$ tensor. In order for the $\chi^{(2)}$ tensor not to vanish, the crystal must not possess inversion symmetry. This is also one of the requirements for a crystal to be piezoelectric. Thus piezoelectric crystals, such as quartz and KDP, are also useful for second harmonic generation of light.

Consider a plane wave of angular frequency ω propagating through a crystal that has the necessary type of symmetry to produce the second harmonic 2ω. The electromagnetic field of the fundamental wave has a space–time variation $e^{i(k_1 z - \omega t)}$, whereas that of the second harmonic is $e^{i(k_2 z - 2\omega t)}$.

Suppose that the crystal is in the form of a slab of thickness l. Then the amplitude of the second harmonic at the exit face of the crystal is obtained by adding the contributions from each element of thickness dz within the crystal, namely,

$$E(2\omega,l) \propto \int_0^l E^2(\omega,z)\, dz$$
$$\propto \int_0^l e^{2i[k_1 z - \omega(t - \tau)]}\, dz \qquad (6.154)$$

Here τ is the time for the optical disturbance of frequency 2ω to travel from z to l. It is given by

$$\tau = \frac{k_2(l - z)}{2\omega} \qquad (6.155)$$

On performing the integration and taking the square of the absolute value, one finds the intensity of the second harmonic to be

$$|E(2\omega)|^2 \propto \left[\frac{\sin (k_1 - \tfrac{1}{2}k_2)l}{k_1 - \tfrac{1}{2}k_2} \right]^2 \qquad (6.156)$$

The above result shows that if $k_1 = \tfrac{1}{2}k_2$, the intensity of second harmonic light is proportional to the square of the slab thickness. Otherwise the maximum intensity is that which can be obtained with a crystal of thickness

$$l_c = \frac{\pi}{2k_1 - k_2} \qquad (6.157)$$

This is known as the "interaction length." Due to dispersion, the interaction length is only $10\lambda_0$ to $20\lambda_0$ for typical crystals. However, it is possible to increase it greatly by the method of *velocity matching*. In this method one makes use of the double nature of the k surface or the velocity surfaces of doubly refracting crystals. Actually, since the energy travels along the ray, it is the ray-velocity surface that is important in this application. Consider a uniaxial crystal. By a suitable choice of the ray direction, it is possible to have the ray velocity of the fundamental (corresponding to an ordinary ray) equal that of the second harmonic (corresponding to the extraordinary ray). This is illustrated in Figures 6.25 and 6.26. With velocity matching, the efficiency for second harmonic generation of light in crystals can be improved by several orders of magnitude.

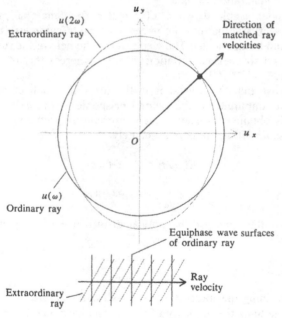

Figure 6.25. Use of the ray-velocity surfaces for velocity matching in the generation of optical harmonics.

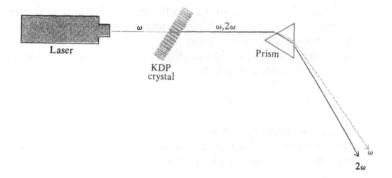

Figure 6.26. Simplified diagram showing the arrangement for optical frequency doubling. The KDP crystal is oriented at the velocity-matched angle.

PROBLEMS

6.1 Show that if the imaginary part κ of the complex index of refraction is much smaller than the real part n, then for the case of a single resonance frequency ω_0, the following approximate equations are valid:

$$n = 1 + \frac{Ne^2}{2m\epsilon_0}\left(\frac{1}{\omega_0{}^2 - \omega^2}\right)$$

$$\kappa = \frac{Ne^2}{2m\epsilon_0}\left(\frac{\gamma\omega}{(\omega_0{}^2 - \omega^2)^2}\right)$$

6.2 Using the above result, show that the maximum and minimum values of n occur at frequencies that define the half-maximum values of κ (see Figure 6.1).

6.3 Derive Sellmeier's semiempirical equation for the index of refraction of a nonabsorbing medium as a function of wavelength:

$$n^2 = 1 + \frac{A_1\lambda^2}{\lambda^2 - \lambda_1{}^2} + \frac{A_2\lambda^2}{\lambda^2 - \lambda_2{}^2} + \cdots$$

6.4 A hypothetical metal has a plasma frequency $\omega_p = 10^{15}$ s^{-1} and a relaxation time $\tau = 10^{-13}$ s. Find the real and imaginary parts of the index of refraction at $\omega = \omega_p$, $\omega = 2\omega_p$, and $\omega = \omega_p/2$.

6.5 The reflectance of a metal is 80 percent at normal incidence, and the extinction index $\kappa = 4$. Find the real part n of the complex index of refraction.

6.6 The conductivity of silver is 6.8×10^7 mho/m. Assuming that the charge carriers are free electrons of density $1.5 \times 10^{28}/\text{m}^3$, find the following quantities: (a) the plasma frequency, (b) the relaxation time, (c) the real and imaginary parts of the index of refraction, and (d) the reflectance at $\lambda = 1 \ \mu$.

6.7 Show that the phase change that takes place on reflection at normal incidence is equal to

$$\tan^{-1} \left(\frac{2\kappa}{n^2 + \kappa^2 - 1} \right)$$

(Note: In using the above result it is necessary to take the correct branch of the arctan function. This is done by requiring that, as $\kappa \to 0$, the phase change is π for $n > 1$, and 0 for $n < 1$.)

6.8 For aluminum at $\lambda = 500$ nm, $n = 1.5$, and $\kappa = 3.2$. Find the normal reflectance and the phase change on reflection.

6.9 Fill in the steps leading to the coefficient of reflection for TM polarization at the boundary of an absorbing medium

$$r_p = \frac{-\mathcal{N} \cos \theta + \cos \phi}{\mathcal{N} \cos \theta + \cos \phi}$$

as discussed in Section 6.6.

6.10 Show that if one introduces an *effective* index of refraction

$$n_{\text{eff}} = \frac{\sin \theta}{\sin \phi}$$

then, for an absorbing medium whose complex index of refraction is $\mathcal{N} = n + i\kappa$, the following equation holds for oblique incidence:

$$(n_{\text{eff}}^2 - n^2 + \kappa^2)(n_{\text{eff}}^2 - \sin^2 \theta) = n^2 \kappa^2$$

6.11 A uniaxial crystal of indices n_O and n_E is cut so that the optic axis is perpendicular to the surface. Show that for a light ray incident from the outside at an angle of incidence θ, the angle of refraction ϕ_E of the extraordinary ray is given by

$$\tan \phi_E = \frac{n_O}{n_E} \frac{\sin \theta}{\sqrt{n_E^2 - \sin^2 \theta}}$$

6.12 Derive Fresnel's equation for the phase-velocity surface:

$$\frac{v_x^2}{c^2 - c^2/n_1^2} + \frac{v_y^2}{v^2 - c^2/n_2^2} + \frac{v_z^2}{v^2 - c^2/n_3^2} = 0$$

6.13 A polarizing prism of the Glan type is to be made of quartz. Determine the angle at which the diagonal face should be cut.

6.14 A 30-degree prism is made of quartz. The optic axis is parallel to the apex edge of the prism. A beam of light ($\lambda_{\text{vac}} = 5890$ Å) is incident so that the deviation is approximately minimum. Determine the angle between the E ray and the O ray.

6.15 A Fresnel prism is made of quartz as shown in Figure 6.18. The component prisms are 70-20-90 degrees. Determine the angle between the emerging right and left polarized rays for sodium light.

6.16 Referring to Equation (6.133), determine the relationship between the value of χ_{12} and the specific rotatory power of quartz.

6.17 A beam of linearly polarized light is sent through a piece of solid glass tubing 25 cm long and 1 cm in diameter. The tubing is wound with a single layer of 250 turns of enameled copper wire along its entire length. If the Verdet constant of the glass is 0.05 min/Oe/cm, what is the amount of rotation of the plane of polarization of the light when a current of 5 amperes (A) is flowing through the wire? (For simplicity assume that the magnetic field is uniform, that is, neglect end effects.)

CHAPTER 7

Thermal Radiation and Light Quanta

7.1 Thermal Radiation

The electromagnetic energy that is emitted from the surface of a heated body is called *thermal radiation*. This radiation consists of a continuous spectrum of frequencies extending over a wide range. The spectral distribution and the amount of energy radiated depend chiefly on the temperature of the emitting surface.

Regarding the spectral distribution, careful measurements show that at a given temperature there is a definite frequency (or wavelength) at which the radiated power is maximum, although this maximum is very broad. Furthermore, the frequency of the maximum is found to vary in direct proportion to the absolute temperature. This rule is known as *Wien's law*. At room temperature, for example, the maximum occurs in the far infrared region of the spectrum, and there is no perceptible visible radiation emitted. But at higher temperatures the maximum shifts to correspondingly higher frequencies. Thus at about 500°C and above, a body glows visibly.

The rate at which energy is radiated by a heated body is also found to have a definite temperature dependence. Measurements indicate that the total power increases as the *fourth power* of the absolute temperature. This is known as the *Stefan-Boltzmann law*. Both the Stefan-Boltzmann law and Wien's law may be regarded as empirical statements about thermal radiation. It is the purpose of the present chapter to derive these laws from basic theory and, in so doing, to deduce other quantitative relationships that apply to radiation emitted by heated bodies.

7.2 Kirchhoff's Law. Blackbody Radiation

Consider a hypothetical situation in which a hollow cavity contains inside it a single body thermally insulated from the cavity walls, say by a nonconducting thread (Figure 7.1). Let the cavity walls be maintained at a constant temperature. Thermal radiation then fills the

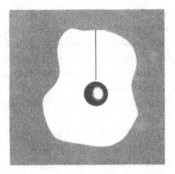

Figure 7.1.   A body inside a cavity. The body exchanges heat with the cavity by thermal radiation and comes to equilibrium when the temperature of the body is the same as the temperature of the cavity walls.

cavity and some of it is absorbed by the body. The body also emits thermal radiation, and equilibrium is attained when the body emits at the same rate that it absorbs. The temperature of the body is then equal to the temperature of the cavity walls.

Let I denote the *irradiance* of the thermal radiation within the cavity; that is, I is the total power per unit area incident on the body. Let b be the fraction of incident power that the body absorbs, and call H the *radiance* or power per unit area that it emits. Then, for equilibrium,

$$H = bI \qquad (7.1)$$

Now suppose that instead of just one body, there are several bodies of different bs inside the cavity. Distinguishing these by subscripts 1, 2, . . . , then, for thermal equilibrium of each body, we have $H_1 = b_1 I$, $H_2 = b_2 I$, . . . , and so forth. It follows that

$$I = \frac{H_1}{b_1} = \frac{H_2}{b_2} = \cdots \qquad (7.2)$$

Thus at a given temperature the ratio of the emitted power to the fraction of power absorbed is the same for all bodies, and is equal to the irradiance within a hollow cavity. This rule is known as *Kirchhoff's law*. According to it, good absorbers are also good emitters and conversely. This fact is easily demonstrated by placing a small spot of lampblack on a glass rod. If the rod is heated to incandescence, the blackened spot appears much brighter than the rest of the rod.

A perfect absorber is called a *blackbody*. For such a body $b = 1$, and the corresponding value of H is the maximum possible, namely,

$$H_{\max} = I \qquad (7.3)$$

Thus a blackbody is the most efficient emitter of thermal radiation, and the emitted power per unit area is equal to the irradiance within a hollow cavity. For this reason blackbody radiation is also known as cavity or *hohlraumstralung*. A practical blackbody radiator can be produced by merely piercing a small hole in an otherwise closed cavity. If the walls of the cavity are maintained at a given temperature, the thermal radiation streaming through the hole is essentially identical with that of a perfect blackbody.

We now proceed to calculate the rate at which radiation is emitted through a hole in a cavity. Let us call u the energy density of all frequencies of the thermal radiation inside the cavity. The *spectral density* u_ν is defined as the energy density per unit frequency interval centered at a given frequency ν. In terms of spectral density, we have

$$u = \int_0^\infty u_\nu \, d\nu \qquad (7.4)$$

This radiation streams about with speed c in all directions. Thus the fraction $d\Omega/4\pi$ of the radiation can be considered to be going in any one direction specified to within an element of solid angle $d\Omega$. Consider a hole of unit area. In one unit of time, an amount of energy $uc \cos \theta \, d\Omega/4\pi$ streams through the hole, where θ is the angle between the direction of the radiation and the normal to the plane of the hole (Figure 7.2).

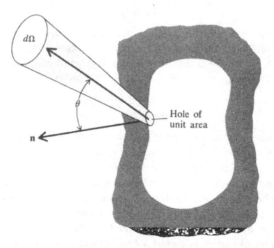

Figure 7.2.   Radiation streaming through a hole in a cavity. The unit vector **n** is normal to the hole and the radiation is spread over a solid angle $d\Omega$.

We now calculate the total amount of radiant energy passing through the hole in all possible directions within a solid angle of 2π, that is, into a hemisphere. The element of solid angle is $d\Omega = \sin\theta \, d\theta \, d\phi$, and the limits of integration are $\theta = 0$ to $\pi/2$, and $\phi = 0$ to 2π. Thus

$$\int_0^{2\pi} \int_0^{\pi/2} uc \cos\theta \sin\theta \, \frac{d\theta \, d\phi}{4\pi} = \frac{uc}{4}$$

is the total radiation emitted per unit time per unit area. Hence

$$I = \frac{uc}{4} \tag{7.5}$$

is the radiance of a blackbody. Corresponding to the spectral energy density u_ν, the spectral radiance is

$$I_\nu = \frac{u_\nu c}{4} \tag{7.6}$$

which is the radiated power per unit area per unit frequency interval centered at frequency ν.

The term *intensity,* denoted by $\mathscr{I}$, is sometimes used to express the power per unit solid angle per unit area. For radiation in a direction normal to the surface

$$\mathscr{I} = \frac{uc}{4\pi} = \frac{I}{\pi} \tag{7.7}$$

The *spectral intensity* is likewise given by

$$\mathscr{I}_\nu = \frac{u_\nu c}{4\pi} = \frac{I_\nu}{\pi} \tag{7.8}$$

which is the power per unit area per unit solid angle per unit frequency interval in a direction normal to the surface.

7.3 Modes of Electromagnetic Radiation in a Cavity

In order to find the density of radiation inside a cavity we must first investigate the standing wave patterns or *modes* of electromagnetic radiation that can exist in it. We shall find that the number of such modes in a given frequency range is of central importance to the theory of radiation. For simplicity we consider a cavity of rectangular shape. Standing waves in the cavity can then be represented by suitable linear combinations of wave functions based on the fundamental wave function

$$e^{i(\mathbf{k}\cdot\mathbf{r}-\omega t)} = e^{ik_x x} \, e^{ik_y y} \, e^{ik_z z} \, e^{-i\omega t}$$

where k_x, k_y, and k_z are the components of **k**. Let A, B, and C be the linear dimensions of the cavity in the x, y, and z directions, respectively. Then a stationary pattern or *mode* will exist if the wave function is periodic in a manner expressed by the following equations:

$$k_x A = \pi n_x \qquad k_y B = \pi n_y \qquad k_z C = \pi n_z \tag{7.9}$$

where n_x, n_y, and n_z are integers. Each set (n_x, n_y, n_z) corresponds to a possible mode of the radiation in the cavity (Figure 7.3). Since $k^2 = k_x^2 + k_y^2 + k_z^2$, then

$$k^2 = \frac{\omega^2}{c^2} = \pi^2 \left(\frac{n_x^2}{A^2} + \frac{n_y^2}{B^2} + \frac{n_z^2}{C^2} \right) \tag{7.10}$$

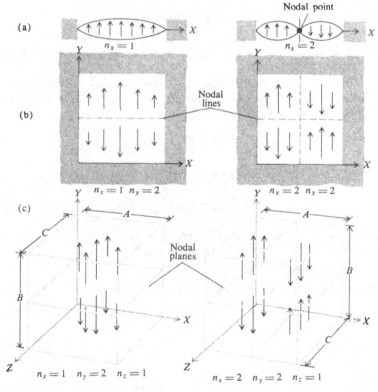

Figure 7.3. Standing wave patterns (modes) in various cavities. Part (a) shows the two lowest modes ($n = 1$ and $n = 2$) of a one-dimensional cavity. In (b) are shown the (1,2) and (2,2) modes of a two-dimensional cavity. Finally, in (c) the (1,2,1) and (2,2,1) modes of a three-dimensional cavity are illustrated.

or, equivalently,

$$\frac{4\nu^2}{c^2} = \frac{n_x^2}{A^2} + \frac{n_y^2}{B^2} + \frac{n_z^2}{C^2} \tag{7.11}$$

The above result shows that for a given frequency ν, only certain values of n_x, n_y, and n_z are allowed.

Let us examine Figure 7.4 in which Equation (7.11) is plotted

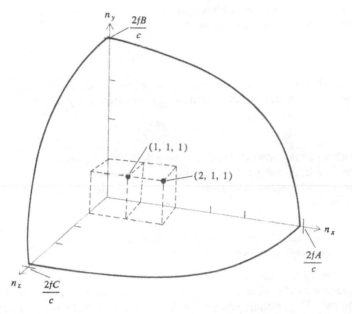

Figure 7.4.   Unit cubes and associated points representing modes of a cavity. One octant of the limiting ellipsoid is shown.

graphically in terms of coordinates n_x, n_y, and n_z. The various modes are represented here by points at the corners of unit cubes, some of which are indicated in the figure. Now Equation (7.11) is the equation of an ellipsoid whose semiaxes are given by $2\nu A/c$, $2\nu B/c$, and $2\nu C/c$. The volume of one octant of this ellipsoid is therefore

$$\frac{1}{8} \frac{4\pi}{3} \frac{2\nu A}{c} \frac{2\nu B}{c} \frac{2\nu C}{c} = \frac{4\pi\nu^3 ABC}{3c^3} = \frac{4\pi\nu^3}{3c^3} V \tag{7.12}$$

where $V = ABC$ is the volume of the cavity. Since each unit cube is associated with one mode, the above expression is equal to the number of modes for *all frequencies equal to or less* than ν. Only one octant of the ellipsoid is needed to count the modes, because both positive and negative values of the ns correspond to the same mode.

It is necessary, however, to double the above value to obtain the total number of modes, since for a given direction of propagation there are two orthogonal polarizations of the electromagnetic radiation inside the cavity. The final value for the number of modes g per unit volume, for all frequencies equal to or less than ν, is

$$g = \frac{8\pi}{3c^3} \nu^3 \tag{7.13}$$

We can now find, by differentiation, the number of modes per unit volume for frequencies lying between ν and $\nu + d\nu$. The result is

$$dg = \frac{8\pi}{c^3} \nu^2 \, d\nu \tag{7.14}$$

A convenient way of interpreting the above result is to say that the number of modes per unit volume per unit frequency interval is

$$g_\nu = \frac{8\pi\nu^2}{c^3} \tag{7.15}$$

Although the above formula has been derived for a rectangular cavity, the result is independent of the shape of the cavity provided the cavity dimensions are large compared to the wavelength of the radiation.

7.4 Classical Theory of Blackbody Radiation. The Rayleigh-Jeans Formula

According to classical kinetic theory, the temperature of a gas is a measure of the mean thermal energy of the molecules that comprise the gas. The average energy associated with each degree of freedom of a molecule is $\frac{1}{2}kT$, where k is Boltzmann's constant and T is the absolute temperature. This well-known rule is called the *principle of equipartition of energy*. It applies, of course, only to systems in thermodynamic equilibrium.

Lord Rayleigh and Sir James Jeans suggested that the equipartition principle might also apply to the electromagnetic radiation in a cavity. If the radiation is in thermal equilibrium with the cavity walls, then one might reasonably expect an equipartition of energy among the cavity modes. Rayleigh and Jeans assumed that the mean energy per mode is kT. In effect, this assumption amounts to saying that in a given mode the electric field and the magnetic field each represent one degree of freedom. If there are g_ν modes per unit frequency interval per unit volume, then the spectral density of the radiation would be $g_\nu kT$. Hence, from Equation (7.15), we have

$$u_\nu = g_\nu kT = \frac{8\pi\nu^2 kT}{c^3} \tag{7.16}$$

This yields, in view of Equation (7.7), the following formula for the spectral radiance, that is, the power per unit area per unit frequency interval:

$$I_\nu = \frac{2\pi\nu^2 kT}{c^2} \tag{7.17}$$

This is the famous Rayleigh-Jeans formula. It predicts a frequency-squared dependence for the spectral distribution of blackbody radiation (Figure 7.5). For sufficiently low frequencies the for-

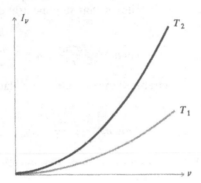

Figure 7.5. The Rayleigh-Jeans law. The curves of I_ν versus ν are parabolas.

mula is found to agree quite well with experimental data. However, at higher and higher frequencies the formula predicts that a blackbody will emit more and more radiation. This, of course, is in contradiction with observation. It is the so-called "ultraviolet catastrophe" of the classical radiation theory. The ultraviolet catastrophe clearly shows that there is a fundamental error in the classical approach.

7.5 Quantization of Cavity Radiation

The way to avert the ultraviolet catastrophe was discovered by Planck in 1901. By introducing a radical concept, namely, the *quantization* of electromagnetic radiation, Planck was able to derive an equation for blackbody emission that was in complete accord with experimental observations. This marked the beginning of quantum theory.

Planck did not accept the principle of equipartition for cavity radiation. He assumed that the energy associated with each mode was quantized; that is, the energy could exist only in integral multiples of some lowest amount or *quantum*. He postulated that the energy of the quantum was proportional to the frequency of the radiation. The name given to the quantum of electromagnetic radiation is the *photon*.

If we call the constant of proportionality h, then $h\nu$ is the energy of a photon of frequency ν. According to Planck's hypothesis, the modes of a cavity are occupied by integral numbers of photons and accordingly the energy of a given mode of frequency ν can have any of the values

$$0, \; h\nu, \; 2h\nu, \; 3h\nu, \; \cdots$$

Let us denote the *average* number of photons per mode by $\langle n_\nu \rangle$. This number is known as the *occupation index*. In terms of it the mean energy per mode is $h\nu \langle n_\nu \rangle$. In view of Equation (7.15), which gives the mode density, it follows that the spectral density of cavity radiation is given by

$$u_\nu = g_\nu h\nu \langle n_\nu \rangle = \frac{8\pi h\nu^3}{c^3} \langle n_\nu \rangle \qquad (7.18)$$

The corresponding spectral radiance function for blackbody radiation is therefore

$$I_\nu = \frac{1}{4} c g_\nu h\nu \langle n_\nu \rangle = \frac{2\pi h\nu^3}{c^2} \langle n_\nu \rangle \qquad (7.19)$$

Our next task is to determine how the occupation index $\langle n_\nu \rangle$ varies as a function of frequency.

7.6 Photon Statistics. Planck's Formula

The problem of finding the particular way that photons are distributed among the available modes of a cavity is an exercise in statistics. We address ourselves to the following question: Given a certain amount of energy in the form of photons in a cavity, is there a particular distribution that is more likely to occur than any other distribution, and if so, what is this most probable distribution? The answer is found by using a well-known method of statistical mechanics used to calculate distribution functions for systems containing large numbers of particles. In this method one calculates the total number of ways W in which the particles, in this case photons, can be arranged in any arbitrary distribution that satisfies certain general conditions. One then proceeds to find that particular distribution for which W is a maximum. Owing to the fact that the total number of particles is extremely large for the cases of interest, the distribution having the largest value of W turns out to be overwhelmingly more probable than any other, and thus represents the actual distribution with virtually absolute certainty.

To apply the statistical method to cavity radiation, we shall divide the frequency spectrum up into an infinite number of intervals. The size of the intervals is arbitrary. For convenience we choose them to be unit intervals. The number of available quantum states (modes) in each interval is g_ν. Let N_ν be the number of photons in an interval; that is, N_ν is the number of photons per unit frequency range centered at frequency ν. The occupation index is then that value of N_ν/g_ν which maximizes W, namely,

$$\langle n_\nu \rangle = \left(\frac{N_\nu}{g_\nu}\right)_{\text{max}} \tag{7.20}$$

Now to find the number of arrangements of the N_ν photons among the g_ν different modes of the interval in question, we can think of the photons as identical objects placed in a linear array of g_ν compartments (Figure 7.6). The photons are represented as dots, and the par-

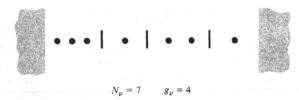

$$N_\nu = 7 \qquad g_\nu = 4$$

Figure 7.6.   One possible arrangement of seven identical objects (photons) in four compartments (modes).

titions between the compartments are represented as vertical lines of which there are $g_\nu - 1$. The total number of ways of arranging the dots in the compartments is just the total number of permutations of all of these $N_\nu + g_\nu - 1$ objects. This is $(N_\nu + g_\nu - 1)!$. But the dots are identical objects, so we must divide by the number of permutations of the dots: $N_\nu!$. Similarly, the partitions are identical, so we must also divide by the number of permutations of the partitions: $(g_\nu - 1)!$. The result is

$$W_\nu = \frac{(N_\nu + g_\nu - 1)!}{N_\nu! \, (g_\nu - 1)!} \tag{7.21}$$

This is the number of different ways of placing N_ν identical objects in g_ν compartments, that is, the number of different arrangements of the N_ν photons in a unit frequency interval containing g_ν modes. Finally, the total number of ways W of arranging the photons in all of the intervals is given by taking the product of all the W_ν, namely,

$$W = \prod_v W_v = \prod_v \frac{(N_\nu + g_\nu - 1)!}{N_\nu! \, (g_\nu - 1)!} \tag{7.22}$$

The factorial function is awkward to handle, so we shall use

Stirling's approximation:

$$\ln x! \cong x \ln x - x \qquad (7.23)$$

This is not very accurate for small x, but it becomes increasingly accurate the larger x becomes.[1] In our application, x is very large. Thus

$$\ln W = \sum_\nu [(N_\nu + g_\nu - 1) \ln (N_\nu + g_\nu - 1)$$
$$- N_\nu \ln N_\nu - (g_\nu - 1) \ln (g_\nu - 1)] \qquad (7.24)$$

If the distribution is such that W is maximum, then $\ln W$ is also maximum, and the first variation $\delta(\ln W)$ is zero. (Actually this is the condition for an extreme value of W. It can be shown, however, that our final result is, in fact, a maximum.) Thus for a maximum we must have

$$\delta(\ln W) = \sum_\nu [\ln (N_\nu + g_\nu) - \ln N_\nu]\delta N_\nu = 0 \qquad (7.25)$$

Here we have neglected unity in comparison to $N_\nu + g_\nu$, which is presumably much greater than unity. Now if the N_νs were all independent quantities, then each separate bracket in the above summation would necessarily have to vanish in order that the equation be valid. However, the N_νs are not actually independent. This stems from the fact that the total photon energy $E = \Sigma h\nu N_\nu$ remains constant. Consequently the variation in the total energy must be zero, namely

$$\delta E = \sum_\nu h\nu \delta N_\nu = 0 \qquad (7.26)$$

To find N_ν as a function of ν such that both Equations (7.25) and (7.26) are simultaneously satisfied, we use Lagrange's method of undetermined multipliers. This method is essentially a way of combining the two equations to obtain a single equation in which the N_νs are effectively independent. Multiply the conditional Equation (7.26) by the undetermined multiplier, a constant that we shall call $-\beta$. Add this to the first equation. The result is

$$\delta (\ln W) - \beta \delta E = 0$$

which gives

$$\sum_\nu [\ln (N_\nu + g_\nu) - \ln N_\nu - \beta h\nu] \delta N_\nu = 0 \qquad (7.27)$$

---

[1] An asymptotic expansion of $\ln x!$ is the following:

$$\ln x! = x\ln x - x + \ln\sqrt{2\pi x} + \ln \left(1 + \frac{1}{12x} + \frac{1}{288 x^2} - \frac{139}{51840 x^3} + \cdots\right)$$

Since $x \gg 1$, the first two terms are the only ones of importance for our purposes.

We now choose β such that each bracket vanishes in the above summation, namely,

$$\ln (N_\nu + g_\nu) - \ln N_\nu - \beta h\nu = 0 \qquad (7.28)$$

On solving for N_ν/g_ν we find the following result for the occupation index

$$\langle n_\nu \rangle = \left(\frac{N_\nu}{g_\nu}\right)_{\text{max}} = \frac{1}{e^{\beta h\nu} - 1} \qquad (7.29)$$

This is the particular distribution that maximizes W subject to the condition that E is constant. It is known as the *Bose-Einstein distribution law for photons*. Other particles besides photons obey a similar law. These particles are collectively known as *bosons*. Examples are alpha particles, pi mesons, and so forth. (A different kind of statistical behavior is exhibited by a second class of particles. The distribution law for this class is called *Fermi-Dirac stastics,* and the particles are known as *fermions*. Electrons, protons, and mu mesons are examples of fermions.) For a full discussion of this subject the reader should consult any standard text on quantum statistics.

Equation (7.29) gives the average number of photons per mode as a function of frequency and the, as yet, unknown constant β. A graphical plot is shown in Figure 7.7.

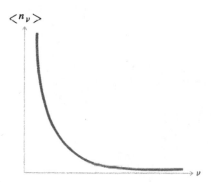

Figure 7.7. Number of photons per unit frequency interval (occupation number) as a function of frequency for one temperature.

Substituting the above expression in Equation (7.19) we find

$$I_\nu = \frac{2\pi h\nu^3}{c^2} \frac{1}{e^{\beta h\nu} - 1} \qquad (7.30)$$

giving the distribution of blackbody radiation as a function of frequency.

For small frequencies ($\beta h_\nu \ll 1$) the above formula reduces to

$$I_\nu = \frac{2\pi\nu^2}{c^2}\frac{1}{\beta} \tag{7.31}$$

This is identical with the Rayleigh-Jeans formula, *provided* we identify the undetermined multiplier β with the quantity $1/kT$. This we shall do on the basis of the physical fact that the Rayleigh-Jeans formula is in agreement with the experimentally observed intensity distribution for low frequencies.

Our final formula for the spectral distribution of blackbody radiation is then

$$I_\nu = \frac{2\pi h\nu^3}{c^2}\frac{1}{e^{h\nu/kT} - 1} \tag{7.32}$$

which is the famous equation first derived by Planck. The equation is in complete agreement with experimental measurements. Figure 7.8(a) shows some curves of I_ν plotted as a function of frequency for

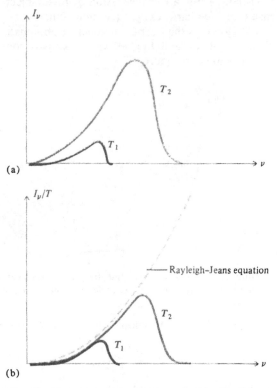

(a)

(b)

Figure 7.8. The Planck radiation law. (a) I_ν versus ν, (b) I_ν/T versus ν with the Rayleigh-Jeans law (dotted) for comparison.

various temperatures. The same data are shown in Figure 7.8(b) except that the quantity I_ν/T is plotted as a function of frequency in order to bring out more closely the comparison between the quantum formula (7.32) and the classical Rayleigh-Jeans formula (7.17).

Both Wien's law and the Stefan-Boltzmann law are readily deduced from the Planck radiation formula. Let us introduce the dimensionless parameter

$$x = \frac{h\nu}{kT} \tag{7.33}$$

Then Planck's formula may be written

$$I_\nu = \frac{2\pi k^3 T^3}{c^2 h^2} \frac{x^3}{e^x - 1} \tag{7.34}$$

By differentiating with respect to x and setting the result equal to zero, it is found that I_ν is maximum for $x \approx 2.82$. This means that the frequency ν_{max} at which I_ν is maximum is given by

$$\nu_{max} = \frac{2.82 \, kT}{h} \tag{7.35}$$

This is one form of Wien's law. (Another way of stating Wien's law is in terms of wavelength. See Problem 7.8 at the end of the chapter.)

To find the total radiation integrated over all frequencies, we have

$$I = \int_0^\infty I_\nu \, d\nu = \frac{2\pi k^4 T^4}{c^2 h^3} \int_0^\infty \frac{x^3 \, dx}{e^x - 1} \tag{7.36}$$

The value of the definite integral is $\pi^4/15$. Thus we obtain the Stefan-Boltzmann law

$$I = \sigma T^4 \tag{7.37}$$

where the Stefan-Boltzmann constant σ is given by

$$\sigma = \frac{2\pi^5 k^4}{15 c^2 h^3} \tag{7.38}$$

The numerical value of the Stefan-Boltzmann constant in MKS units is 5.67×10^{-8} watts/m^2/degree^4. Thus a blackbody radiator at a temperature of 1000°K will emit at a rate of 56.7 Kw per square meter of surface area.

7.7 The Photoelectric Effect and the Detection of Individual Photons

When a metal surface is exposed to a beam of light, electrons are emitted from the surface of the metal. This well-known phenomenon, the photoelectric effect, is used in many ways for the measurement and control of light, and so forth.

The photoelectrons are emitted with a wide range of kinetic energies. Careful measurements show that if the light is monochromatic, then the photoelectrons, although having a certain energy spread, never have energies greater than a certain amount E_{max}. The value of E_{max} is found to depend linearly on the frequency ν of the incident light according to the following equation:

$$E_{max} = h\nu - e\phi \qquad (7.39)$$

where h is Planck's constant and e is the electronic charge. The value of the constant ϕ depends on the particular metal used and is known as the *work function* of the metal. It is of the order of a few volts for most metals. The quantity $e\phi$ represents essentially the work required to separate the electron from the metal surface.

According to the above equation, the energy of the emitted electron does not depend at all on the intensity of the incident light but only on the frequency. However, the photocurrent is found to vary directly with the intensity of the light. In other words, the number of photoelectrons emitted per second is directly proportional to the number of photons per second striking the metal surface. Now the intensity of the light can be made so low that individual photoelectrons may be counted. Since the energy of each photoelectron is still represented by Equation (7.39), we must conclude that energy is transferred to an electron by a single photon. The photoelectric effect thus makes possible the detection of individual photons.

7.8   Momentum of a Photon. Light Pressure

According to a well-known derivation based on classical electromagnetic theory [16], [34], the pressure exerted by a beam of light of irradiance I on a black surface is I/c. This was demonstrated experimentally many years ago [28]. The magnitude of the effect is very small, however, and requires a very sensitive pressure-measuring device to detect it.

In terms of the quantum description of light, the existence of light pressure implies that photons must carry momentum as well as energy. The linear momentum of a photon can be calculated very easily if it is assumed that Einstein's mass–energy relation holds, that is,

$$h\nu = mc^2 \qquad (7.40)$$

where ν is the frequency and m is the mass of the photon.[2] Since the

---

[2] The rest mass m_0 of a photon must be zero. Otherwise, by the mass–velocity formula

$$m = m_0 \frac{1}{\sqrt{1 - u^2/c^2}}$$

m would be infinite, since $u = c$.

speed of the photon is c, the linear momentum p is given by

$$p = mc = \frac{h\nu}{c} \qquad (7.41)$$

An alternative way of expressing the momentum of a photon is in terms of its wavelength λ. This expression is

$$p = \frac{h}{\lambda} \qquad (7.42)$$

which follows immediately from the fact that $\nu\lambda = c$.

Suppose that a stream of photons falls normally on a perfectly absorbing surface. Assuming that momentum is conserved, a given photon transfers its entire momentum $h\nu/c$ to the surface on being absorbed. If there are N photons per unit area that hit the surface every second, then the pressure P, being equal to the time rate of transfer of linear momentum per unit area, is given by

$$P = \frac{Nh\nu}{c} \qquad (7.43)$$

Now the irradiance I of the beam is the power per unit area. Since each photon carries an energy $h\nu$, it follows that

$$I = Nh\nu \qquad (7.44)$$

Consequently,

$$P = \frac{I}{c} \qquad (7.45)$$

which is the same as the classical expression for the pressure.

If the surface is perfectly reflecting, then the pressure is just *twice* the above value, namely $2I/c$. This is because a photon, on being reflected back, undergoes a change of momentum of $p - (-p) = 2p$. Each photon therefore transfers twice as much momentum to the surface as it would if it had been absorbed.

7.9 Angular Momentum of a Photon

If a beam of circularly polarized light is incident on an absorbing surface, classical electromagnetic theory predicts that the surface must experience a torque [34]. The calculation gives the magnitude $\mathcal{T}$ of the torque per unit area as

$$\mathcal{T} = \frac{I}{2\pi\nu} = \frac{I}{\omega} \qquad (7.46)$$

where I is the irradiance of the beam, as before. The above result, in view of (7.44), can be written in the alternative form

$$\mathscr{T} = \frac{Nh}{2\pi} \qquad (7.47)$$

This implies that photons must carry angular momentum as well as linear momentum. This intrinsic angular momentum is called *spin*. From the above equation we see the magnitude of the spin of a photon has the value $h/2\pi$. For right circularly polarized light, the direction of the spin of the photon is parallel to the direction of propagation, whereas for left circularly polarized light, it is antiparallel to the direction of propagation. Both linearly polarized light and unpolarized light can be considered to be equal mixtures of right and left circular polarizations, so that the average angular momentum is zero. *In the case of linearly polarized light, the mixture is coherent, whereas for unpolarized light, it is incoherent.*

7.10   Wavelength of a Material Particle.
de Broglie's Hypothesis

In 1924 the French physicist Louis de Broglie proposed that just as light exhibits both wavelike and particlelike properties, so material particles might also exhibit a wavelike behavior. In analogy with the expression for a photon's momentum, $p = h/\lambda$, de Broglie suggested that the same relation might apply to *any* particle. Thus a moving particle of momentum $p = mu$ would have an associated wavelength λ given by

$$\lambda = \frac{h}{p} = \frac{h}{mu} \qquad (7.48)$$

where h is Planck's constant.

The validity of de Broglie's bold hypothesis was verified by a famous experiment performed by Davisson and Germer in 1927. They demonstrated that a beam of electrons impinging on a crystal is reflected from the crystal in just the same way that a beam of light is reflected by a diffraction grating. The electrons were preferentially reflected at certain angles. The values of these angles of strong reflection were found to obey the optical grating formula

$$n\lambda = d \sin \theta \qquad (7.49)$$

where n is an integer and d is the grating spacing. For a crystal, d is the spacing between adjacent rows of atoms of the crystal. In Davisson and Germer's experiment, the wavelength found for the electrons agreed precisely with the de Broglie equation.

Since Davisson and Germer's experiment, many others have been performed verifying de Broglie's formula, not only for electrons, but for other particles as well, including protons, neutrons, simple atoms, and molecules. Light and matter both have dual natures. Each exhibits both particlelike and wavelike properties depending on the experimental situation. This schizophrenic behavior of matter and radiation shows that neither the particle model nor the wave model is strictly correct. They are not mutually exclusive, however, but rather, are *complementary* descriptions. Each model emphasizes a certain aspect and each has its own limitations.

7.11 Heisenberg's Uncertainty Principle

One of the most basic and far-reaching concepts of modern physical theory was formulated in 1927 by Werner Heisenberg and is known as the *uncertainty principle*. This principle is concerned with the limit to the precision of our knowledge about physical systems. Specifically, the principle may be stated as follows: If P and Q are two conjugate variables in the sense of classical mechanics, then the simultaneous values of these variables can never be known to a precision greater than that given by

$$\Delta P \, \Delta Q \approx h \qquad (7.50)$$

where h is Planck's constant. Examples of conjugate quantities are energy and time, position and momentum, angle and angular momentum, and so forth.

To illustrate how the uncertainty principle applies to light quanta, suppose one were given the task of determining the precise energy of a photon or, equivalently, the frequency. Suppose further that this measurement had to be made within a definite time interval Δt, say by letting a light shutter be open for this amount of time and then closing it. Now even if the source of light were strictly monochromatic, the fact that the light pulse lasts for a finite time means that there is a resulting frequency spread in the Fourier resolution of the pulse given by

$$\Delta \nu \, \Delta t \approx 1 \qquad (7.51)$$

This can be written

$$\Delta(h\nu) \, \Delta t \approx h \qquad (7.52)$$

that is,

$$\Delta E \, \Delta t \approx h \qquad (7.53)$$

Here the quantity ΔE means the uncertainty in the energy of the photon, corresponding to the uncertainty $\Delta \nu$ of its frequency. Thus the energy of the photon cannot be measured with absolute precision unless the time interval is infinite. Conversely, the knowledge of the time t at which the photon left the shutter cannot be known with certainty unless we are willing to give up all knowledge concerning its frequency or, equivalently, its energy. In any case the product of the two uncertainties ΔE and Δt can never be less than h.

A second application concerns the position of a photon. As in the previous example, let the shutter be open for a time Δt and suppose the released photons are traveling in the x direction. Then

$$\Delta x = c \, \Delta t \approx \frac{c}{\Delta \nu} \qquad (7.54)$$

But from Equation (7.41),

$$\Delta p = \frac{h \, \Delta \nu}{c} \qquad (7.55)$$

Hence

$$\Delta x \, \Delta p \approx h \qquad (7.56)$$

This means that we cannot know the instantaneous values of both the position and momentum of a photon to an arbitrary degree of accuracy. If we know its momentum with absolute accuracy, we know nothing about its location and, conversely, if we know its position accurately, we have no knowledge at all concerning its momentum.

The uncertainty principle brings out the inadequacy of describing a photon as a particle. If we push the description too far by specifying the photon's exact location in space and time, the description loses meaning, because the momentum and the energy of the photon become completely undetermined.

PROBLEMS

7.1 Compute the number of available modes in a cubical box 10 cm on a side, for the following intervals:

(a) A frequency interval of 10^3 Hz centered at a wavelength of 500 nm.

(b) A wavelength interval of 1 Å centered at 5000 nm.

7.2 Determine the number of photons in the box of the above problem for the two intervals given when the temperature of the walls is (a) room temperature 300°K and (b) the sun's temperature 6000°K.

7.3 A 100-W tungsten lamp operates at a temperature of 1800°K. How many photons does it emit per second in the interval 5000 Å to 5001 Å? (Assume the filament emits as a blackbody.)

7.4 Determine the number of photons per mode in a cavity whose temperature is 300°K and 6000°K for the following wavelengths:

(a) 5000 nm, (b) 50 μm, and (c) 5 mm.

7.5 Calculate the total number of photons of *all* frequencies in a cavity of volume V at a temperature T.

7.6 Calculate the total number of photons per unit area per unit time emitted by a blackbody at temperature T.

7.7 Give the Planck radiation law in terms of power per unit area per unit *wavelength* interval. Express it in terms of wavelength.

7.8 Find the wavelength of maximum radiation per unit wavelength interval. The result will be different from the equivalent wavelength of Equation (7.35). Why?

7.9 The linewidth of a helium–neon laser is 10^3 Hz. The operating wavelength is 6328 Å and the power is 1 mW. (a) How many photons are emitted per second? (b) If the output beam is 1 mm in diameter, at what temperature would a blackbody have to be in order to emit the same number of photons from an equal area and over the same frequency interval as the laser?

7.10 Derive an equation giving the pressure exerted on the walls of a cubical cavity at temperature T due to the photons inside. (Assume that one third of the photons are effective in any one direction, say the x direction.)

7.11 From the result of Problem 7.10, find the temperature at which the pressure is equal to 10^{-3} atmosphere.

7.12 A laser beam is focused to a diffraction-limited spot of 1 μm diameter. If the total power of the laser is 10 W, what is the pressure at the focal point?

CHAPTER 8

Optical Spectra

8.1 General Remarks

A *spectrum* may be defined as an ordering of electromagnetic radiation according to frequency, or what amounts to the same thing, an ordering by wavelength. The complete spectrum of a given source comprises all the frequencies that the source emits. Since no single universal frequency-resolving instrument exists, the various regions of the electromagnetic spectrum must be investigated by different methods. The main regions have already been mentioned in Section 1.4.

The so-called "optical" region extends over a wide range from the far infrared on the one end to the far ultraviolet on the other. It includes the visible region as a relatively small portion (Figure 8.1).

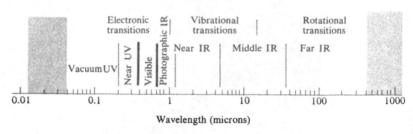

Figure 8.1. The optical region of the electromagnetic spectrum.

In practice the optical region is characterized by (1) the fact that the radiation is focused, directed, and controlled by mirrors and lenses, and (2) the use of prisms and gratings for dispersing the radiation into a spectrum.

Unlike the continuous spectrum of thermal radiation given off by solid bodies, the radiation emitted by excited atoms or molecules is found to consist of various discrete frequencies. These frequencies are characteristic of the particular kinds of atoms or molecules in-

226

volved. The term "line" spectrum is commonly used when referring to such radiation. This terminology originates from the fact that a *slit* is generally the type of entrance aperture for spectroscopic instruments used in the optical region, so a separate *line* image of the slit is formed at the focal plane for each different wavelength comprising the radiation. Sources of line spectra include such things as arcs, sparks, and electric discharges through gases.

The optical spectra of most atoms are quite complex and the line patterns are seemingly random in appearance. A few elements, notably hydrogen and the alkali metals, exhibit relatively simple spectra that are characterized by easily recognized *series* of lines converging toward a limit.

The optical spectra of many molecules, particularly diatomic molecules, appear as more or less regularly spaced "bands" when examined with a spectroscopic instrument of low resolving power. However, under high resolution, these bands are found to be sequences of closely spaced lines.

When white light is sent through an unexcited gas or vapor, it is generally found that the atoms or molecules comprising the vapor absorb just those same frequencies that they would emit if excited. The result is that those particular frequencies are either weakened or are entirely missing from the light that is transmitted through the vapor. This effect is present in ordinary sunlight, the spectrum of which appears as numerous dark lines on a bright continuous background.

The dark lines are called Fraunhofer lines after J. Fraunhofer who made early quantitative measurements. These lines reveal the presence of a relatively cool layer of gas in the sun's upper atmosphere. The atoms in this layer absorb their own characteristic wavelengths from the light coming from the hot, dense surface of the sun below, which emits thermal radiation corresponding to a temperature of about 5500°K.

Selective absorption is also exhibited by solids. Virtually all transparent solids show broad absorption bands in the infrared and ultraviolet. In most colored substances these absorption bands extend into the visible region. However, relatively sharp absorption bands may occur in certain cases such as crystals and glasses that contain rare earth atoms as impurities.

8.2 Elementary Theory of Atomic Spectra

The mathematical theory of atomic spectra had its beginning in 1913 when Niels Bohr, a Danish physicist, announced his now famous work. He was concerned mainly with a theoretical explanation of the spectrum of hydrogen, although his basic ideas are applicable to other systems as well.

In order to explain the fact that atoms emit only certain characteristic frequencies, Bohr introduced two fundamental assumptions. These are

(1) *The electrons of an atom can occupy only certain discrete quantized states or orbits. These states have different energies, and the one of lowest energy is the normal state of the atom, also known as the* ground state.

(2) *When an electron undergoes a transition from one state to another, it can do so by emitting or absorbing radiation. The frequency ν of this radiation is given by*

$$\nu = \frac{\Delta E}{h} \tag{8.1}$$

where ΔE is the energy difference between the two states involved, and h is Planck's constant.

These assumptions represent a radical departure from the classical or Newtonian concept of the atom. The first is suggestive of the quantization of cavity radiation introduced earlier by Planck. The second idea amounts to saying that an atom emits, or absorbs, a single photon upon changing from one quantized state to another, the energy of the photon being equal to the energy difference between the two states (Figure 8.2).

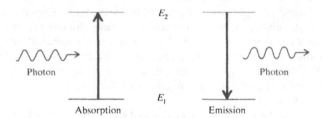

Figure 8.2.   Diagram showing the processes of absorption and emission.

The frequency spectrum of an atom or molecule is given by taking the various possible energy differences $|E_1 - E_2|$ and dividing by h. In order to calculate the actual spectrum of a given atom, one must first know the energies of the various quantum states of the atom in question. Conversely, the energies of the quantum states can be inferred from the measured frequencies of the various spectrum lines.

The Bohr Atom and the Hydrogen Spectrum   In his study of the hydrogen atom, Bohr was able to obtain the correct formula for the energy levels by introducing a fundamental postulate concerning angular momentum. According to this postulate, *the angular momentum of an electron is always an integral multiple of the quantity $h/2\pi$*, where h is Planck's constant.

An electron of mass m traveling with speed u in a circular orbit of radius r has angular momentum mur. Hence the relation

$$mur = \frac{nh}{2\pi} \qquad (n = 1, 2, 3, \ldots) \qquad (8.2)$$

expresses the quantization of the orbital angular momentum of the electron. The integer n is known as the *principal quantum number*.

The classical force equation for an electron of charge $-e$ revolving in a circular orbit of radius r, centered on a proton of charge $+e$, is

$$\frac{e^2}{4\pi\epsilon_0 r^2} = \frac{mu^2}{r} \qquad (8.3)$$

By elimination of u between the two equations, one obtains the following formula for the radii of the quantized orbits:

$$r = \frac{\epsilon_0 h^2}{\pi m e^2} n^2 = a_H n^2 \qquad (8.4)$$

The radius of the smallest orbit ($n = 1$) is called the *first Bohr radius* and is denoted by a_H. Its numerical value is

$$a_H = \frac{\epsilon_0 h^2}{\pi m e^2} = 0.529 \text{ Å} \qquad (8.5)$$

The various orbits are then given by the sequence a_H, $4a_H$, $9a_H$, . . . , and so forth.

The total energy of a given orbit is given by the sum of the kinetic and the potential energies, namely,

$$E = \frac{1}{2} mu^2 - \frac{e^2}{4\pi\epsilon_0 r} \qquad (8.6)$$

Eliminating u by means of Equation (8.3) one finds

$$E = -\frac{e^2}{8\pi\epsilon_0 r} \qquad (8.7)$$

This is the classical value for the energy of a bound electron. If all values of r were allowed, then any (negative) value of energy would be possible. But the orbits are quantized according to Equation (8.4). The resultant quantized energies are given by

$$E_n = -\frac{R}{n^2} \qquad (8.8)$$

in which the constant R, known as the Rydberg constant, is

$$R = \frac{me^4}{8\epsilon_0^2 h^2} \tag{8.9}$$

This is also the binding energy of the electron in the ground state, $n = 1$. Its value in electron volts is approximately 13.5 eV.

The formula for the hydrogen spectrum is obtained by combining the energy equation with the Bohr frequency condition. Calling E_1 and E_2 the energies of the orbits n_1 and n_2, respectively, we find

$$\nu = \frac{E_2 - E_1}{h} = \frac{R}{h} \left(\frac{1}{n_1^2} - \frac{1}{n_2^2} \right) \tag{8.10}$$

The numerical value of R/h is 3.29×10^{15} Hz.[1]

A transition diagram of the hydrogen atom is shown in Figure 8.3. The energies of the various allowed orbits are plotted as horizontal lines, and the transitions, corresponding to the various spectral lines, are shown as vertical arrows. Various combinations of the integers n_1 and n_2 give the observed spectral series. These are as follows:

$n_1 = 1$	$n_2 = 2, 3, 4, \ldots$	Lyman series (far ultraviolet)
$n_1 = 2$	$n_2 = 3, 4, 5, \ldots$	Balmer series (visible and near ultraviolet)
$n_1 = 3$	$n_2 = 4, 5, 6, \ldots$	Paschen series (near infrared)
$n_1 = 4$	$n_2 = 5, 6, 7, \ldots$	Brackett series (infrared)
$n_1 = 5$	$n_2 = 6, 7, 8, \ldots$	Pfund series (infrared)

$\cdot$ $\cdot$ $\cdot$

Some of the series are shown in Figure 8.4 on a logarithmic wavelength scale.

The first three lines of the Balmer series, namely, H_α at a wavelength of 6563 Å, H_β at 4861 Å, and H_γ at 4340 Å, are easily seen by viewing a simple hydrogen discharge tube through a small spectroscope. The members of the series up to $n_2 = 22$ have been recorded by photography. The intensities of the lines of a given series diminish with increasing values of n_2. Furthermore, the intensities of the various series decrease markedly as n_1 increases. Observations using ordinary laboratory sources have extended as far as the line at 12.3 μ in the infrared, corresponding to $n_1 = 6$, $n_2 = 7$.

---

[1] Spectroscopists usually write Equation (8.10) in terms of spectroscopic wavenumber

$$\sigma = \frac{\nu}{c} = \mathscr{R} \, (n_1^{-2} - n_2^{-2})$$

where $\mathscr{R}$ is the Rydberg constant in wavenumber units, namely,

$$\mathscr{R} = \frac{R}{ch} = 10{,}973{,}731 \text{ m}^{-1}$$

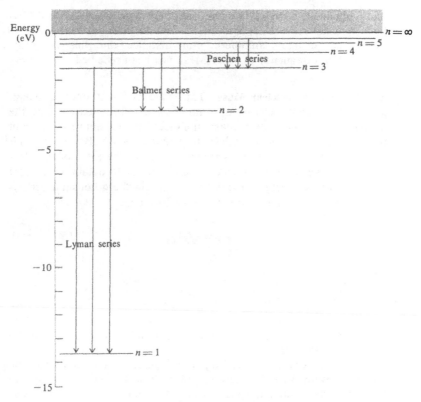

Figure 8.3.   Energy levels of atomic hydrogen. Transitions for the first three series are indicated.

The hydrogen spectrum is of particular astronomical importance. Since hydrogen is the most abundant element in the universe, the spectra of most stars show the Balmer series as prominent absorption lines. The series also appears as bright emission lines in the spectra of

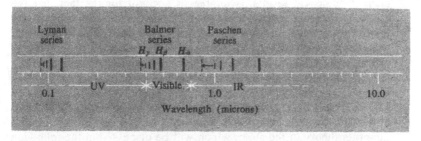

Figure 8.4.   The first three series of atomic hydrogen on a logarithmic wavelength scale.

many luminous nebulas. Recent radiotelescope observations [26] have revealed interstellar hydrogen emission lines corresponding to very large quantum numbers. For instance, the line $n_1 = 158$, $n_2 = 159$ at a frequency of 1651 MHz has been received.

Effect of a Finite Nuclear Mass   The value of the Rydberg constant given by Equation (8.9) is for a nucleus of infinite mass. Since the nucleus actually has a finite mass, the electron does not revolve about the nucleus as a center, rather, both particles revolve about their common center of mass. This requires that m, the mass of the electron, be replaced by the reduced mass in order to obtain the correct value for the Rydberg constant. Thus the Rydberg constant for hydrogen is more accurately given by the formula

$$R_H = \frac{\mu e^4}{8\epsilon_0^2 h^2} \tag{8.11}$$

where

$$\mu = \frac{mM}{m + M} \tag{8.12}$$

is the reduced mass, M is the mass of the atomic nucleus, and m is the mass of the electron.

In the case of ordinary hydrogen the nucleus is a single proton and the mass ratio M/m is equal to 1836. For the heavy isotope of hydrogen, deuterium, the mass ratio is about twice as much. The Rydberg constant for deuterium is therefore slightly different from that for hydrogen. The result is that a small difference exists between the frequencies of corresponding spectrum lines of the two isotopes. This effect, called *isotope shift*, can be seen as a "doubling" of the lines from a discharge tube containing a mixture of hydrogen and deuterium.

The Bohr model of the hydrogen atom, although giving essentially correct numerical results, is unable to account for the fact that the electron does not radiate while traveling in its circular orbit in the ground state, as required by classical electromagnetic theory. Further, the theory is difficult to apply to more complicated atoms and is completely inapplicable to molecules. Early attempts were made by various theorists to modify Bohr's theory in order to account for such things as the fine structure of spectrum lines, and so forth (Section 8.7, below). These attempts met with varying degrees of success. However, the Bohr theory has now been superseded by the modern quantum theory of the atom, which will be discussed in the following sections.

Spectra of the Alkali Metals   An empirical formula similar to that for hydrogen gives fairly accurate results for the spectra of the alkali metals lithium, sodium, and so forth. This formula is

$$\nu = R \left[\frac{1}{(n_1 - \delta_1)^2} - \frac{1}{(n_2 - \delta_2)^2} \right] \tag{8.13}$$

The Rydberg constant R is approximately the same as R_H. It has a slightly different value for each element. The quantum numbers n_1 and n_2 are integers, and the associated quantities δ_1 and δ_2 are known as *quantum defects*.

In a given spectral series, specified by a fixed value of n_1 and a sequence of increasing values of n_2, the quantum defects are very nearly constant. The most prominent series for the alkalis are designated as *sharp, principal, diffuse,* and *fundamental*, respectively. As a typical example, the quantum numbers and quantum defects for the series in sodium are listed in Table 8.1.

Table 8.1. SERIES IN SODIUM

Series	n_1	n_2	δ_1	δ_2
Sharp	3	4, 5, 6, $\cdots$	0.87	1.35
Principal	3	3, 4, 5, $\cdots$	1.35	0.87
Diffuse	3	3, 4, 5, $\cdots$	0.87	0.01
Fundamental	3	4, 5, 6, $\cdots$	0.01	0.00

The sharp and the diffuse series are so named because of the appearance of the spectral lines.

The principal series is the most intense in emission and is also the one giving the strongest absorption lines when white light is passed through the vapor of the metal.

In the case of the fundamental series, the quantum defects are very small. As a consequence the frequencies of the lines of this series are very nearly the same those of the corresponding series in hydrogen. This is the reason for the name "fundamental."

8.3 Quantum Mechanics

Modern quantum theory was pioneered by Schrödinger, Heisenberg, and others in the 1920s. Originally there were two apparently different quantum theories called *wave mechanics* and *matrix mechanics*. These two formulations of quantum theory were later shown to be completely equivalent. Quantum mechanics, as it is known today, includes both. We shall not attempt a rigorous development of quantum mechanics here, but shall merely state some of the essential results that apply to atomic theory.

The quantum mechanical description of an atom or atomic system is made in terms of a *wave function* or *state function*. The commonly used symbol for this function is Ψ. Ordinarily Ψ is a complex number and is considered to be a function of all of the configurational coordinates of the system in question including the time.

According to the basic postulates of quantum mechanics, the state function Ψ has the property that the square of its absolute value, $|\Psi|^2$ or $\Psi^*\Psi$, is a measure of the *probability* that the system in question is located at the configuration corresponding to particular values of the coordinates. $\Psi^*\Psi$ is sometimes referred to as the *probability distribution function* or the *probability density*.

If the system is a single electron, for example, with coordinates x, y, and z, then the probability that the electron is located between x and $x + \Delta x$, y and $y + \Delta y$, z and $z + \Delta z$ is given by the expression

$$\Psi^*(x,y,z,t)\Psi(x,y,z,t)\ \Delta x\ \Delta y\ \Delta z \tag{8.14}$$

It is evident from the above interpretation of the state function that one can never be certain that the electron is located at any given place. Only the chance of its being there within certain limits can be known. This is entirely consistent with the Heisenberg uncertainty principle discussed earlier in Section 7.11.

Now the total probability that the electron is located *somewhere* in space is necessarily unity. It follows that the integral, over all space, of the probability density is finite and has, in fact, the value 1, namely,

$$\int_{-\infty}^{\infty} \int_{-\infty}^{\infty} \int_{-\infty}^{\infty} \Psi^*\Psi\, dx\, dy\, dz = 1 \tag{8.15}$$

Functions satisfying the above equation are said to be *quadratically integrable, normalized* functions.

Stationary States   A *characteristic state* or *eigenstate* is one that corresponds to a perfectly defined energy. A given system may have many eigenstates, each possessing, in general, a different energy. If E_n denotes the particular energy of a system when it is in one of its characteristic states, then the time dependence of the state function is given by the complex exponential factor $\exp(-iE_n t/\hbar)$, where

$$\hbar = \frac{h}{2\pi}$$

Consequently the complete state function is expressible as

$$\Psi_n(x,y,z,t) = \Psi_n(x,y,z)\, e^{-iE_n t/\hbar} \qquad (8.16)$$

Here ψ_n is a function of the configurational coordinates only. It does not involve the time.

Consider the probability density of a system in one of its characteristic states. We have

$$\Psi_n{}^*\Psi_n = \psi_n{}^*e^{iE_n\, t/\hbar}\psi_n e^{-iE_n\, t/\hbar} = \psi_n{}^*\psi_n \qquad (8.17)$$

We see that the exponential factors cancel out. This means that the probability distribution is constant in time, or stationary. Thus characteristic states are also called *stationary states*. A system that is in a stationary state is a static system in the sense that no changes at all are taking place with respect to the external surroundings.

In the particular case in which the quantum-mechanical system is an atom, consisting of a nucleus with surrounding electrons, the probability distribution function is actually a measure of the mean electron density at a given point in space. One sometimes refers to this as a *charge cloud*. When an atom is in a stationary state, the electron density is constant in time. The surrounding electromagnetic field is static and the atom does not radiate.

Coherent States Consider a system that is in the process of changing from one eigenstate ψ_1 to another ψ_2. During the transition the state function is given by a linear combination of the two state functions involved, namely,

$$\Psi = c_1\psi_1\, e^{-iE_1 t/\hbar} + c_2\psi_2\, e^{-iE_2 t/\hbar} \qquad (8.18)$$

Here c_1 and c_2 are parameters whose variation with time is slow in comparison with that of the exponential factors. A state of the above type is known as a *coherent state*. One essential difference between a coherent state and a stationary state is that the energy of a coherent state is not well defined, whereas that of a stationary state is.

The probability distribution of the coherent state represented by Equation (8.18) is given by the following expression:

$$\Psi^*\Psi = c_1{}^*c_1\psi_1{}^*\psi_1 + c_2{}^*c_2\psi_2{}^*\psi_2 + c_1{}^*c_2\psi_1{}^*\psi_2\, e^{i\omega t} + c_2{}^*c_1\psi_2{}^*\psi_1\, e^{-i\omega t}$$
$$(8.19)$$

where

$$\omega = \frac{E_1 - E_2}{\hbar} \qquad (8.20)$$

or, equivalently,

$$\nu = \frac{E_1 - E_2}{h}$$

The above result shows that *the probability density of a coherent state undergoes a sinusoidal oscillation with time. The frequency of this oscillation is precisely that given by the Bohr frequency condition.*

The quantum-mechanical description of a radiating atom may be stated as follows: During the change from one quantum state to another, the probability distribution of the electron becomes coherent and oscillates sinusoidally; this sinusoidal oscillation is accompanied by an oscillating electromagnetic field that constitutes the radiation.

8.4   The Schrödinger Equation

Thus far we have not discussed the question of just how one goes about finding the state functions of a particular physical system. This is one of the basic tasks of quantum theory, and the performance of this task involves the solution of a differential equation known as the *Schrödinger equation.* A simple derivation of this important equation for the case of a single particle proceeds as follows.

Consider any wave function Ψ whose time dependence has the usual sinusoidal variation. Let λ be the wavelength. Then we know that the spatial part ψ of the wave function must obey the standard time-independent wave equation

$$\nabla^2\psi + \left(\frac{2\pi}{\lambda}\right)^2 \psi = 0$$

Now according to de Broglie's hypothesis (Section 7.10), a particle having momentum p has an associated wavelength h/p. Thus a particle would be expected to obey a wave equation of the form

$$\nabla^2\psi + \left(\frac{2\pi p}{h}\right)^2 \psi = 0$$

But a particle of mass m has energy E given by $E = (\frac{1}{2})mu^2 + V$, in which V is the potential energy and u is the speed. Since the linear momentum $p = mu$, then

$$p^2 = 2m(E - V)$$

The wave equation of the particle can therefore be written as

$$\nabla^2\psi + \frac{8\pi^2 m}{h^2}(E - V)\psi = 0 \qquad \text{(8.21)}$$

This is the famous equation first announced by Erwin Schrödinger in 1926. It is a linear, partial differential equation of the second order. The physics involved in the application of the equation essentially amounts to the selection of a potential function $V(x,y,z)$ appropriate to the particular physical system in question. Given the potential function $V(x,y,z)$, the mathematical problem is that of finding the function (or functions) ψ which satisfy the equation.

Not all mathematical solutions of the Schrödinger equation are physically meaningful. In order to represent a real system the function ψ must tend to zero for infinite values of the coordinates in such a way as to be quadratically integrable. This has already been implied by Equation (8.15).

The details of solving partial differential equations of the Schrödinger type are often very involved and complicated, but the results are easily understood. It turns out that the requirement that the solutions be quadratically integrable leads to the result that acceptable solutions can exist only if the energy E has certain definite values. These allowed values of E, called *eigenvalues*, are, in fact, just the characteristic energy levels of the system. The corresponding solutions are called *eigenfunctions*. They are the state functions of the system.

The Schrödinger equation thus leads to the determination of the energy states of the system as well as the associated state functions. In the next section it is shown how the Schrödinger equation is applied to the problem of calculating the energy levels and state functions of the hydrogen atom.

8.5 Quantum Mechanics of the Hydrogen Atom

The quantum theory of a single electron moving in a central field, briefly outlined here, forms the basis of modern atomic theory. In the mathematical treatment of the one-electron atom it is convenient to employ polar coordinates r, θ, and ϕ, owing to spherical symmetry of the field in which the electron moves. The Laplace operator in these coordinates is

$$\nabla^2 = \frac{1}{r^2}\left[\frac{\partial}{\partial r}\left(r^2\frac{\partial}{\partial r}\right) + \frac{1}{\sin\theta}\frac{\partial}{\partial\theta}\left(\sin\theta\frac{\partial}{\partial\theta}\right) + \frac{1}{\sin^2\theta}\frac{\partial^2}{\partial\phi^2}\right] \quad (8.22)$$

The corresponding Schrödinger equation is

$$\frac{1}{r^2}\left[\frac{\partial}{\partial r}\left(r^2\frac{\partial}{\partial r}\right) + \frac{1}{\sin\theta}\frac{\partial}{\partial\theta}\left(\sin\theta\frac{\partial}{\partial\theta}\right) + \right.$$
$$\left. \frac{1}{\sin^2\theta}\frac{\partial^2}{\partial\phi^2}\right]\psi + \frac{8\pi^2\mu}{h^2}(E - V)\psi = 0 \quad (8.23)$$

Here μ is the reduced mass of the electron. In the case of the hydrogen atom, the potential V is given by

$$V = -\frac{e^2}{4\pi\epsilon_0 r} \quad (8.24)$$

Ground State of the Hydrogen Atom   We shall first obtain a simple solution of the Schrödinger equation by the trial method. Substituting a simple exponential trial solution of the form

$$\psi = e^{-\alpha r} \tag{8.25}$$

where α is an undetermined constant, we find that the Schrödinger equation reduces to

$$\left(\alpha^2 + \frac{8\pi^2\mu E}{h^2}\right) e^{-\alpha r} + \left(\frac{2\pi\mu e^2}{\epsilon_0 h^2} - 2\alpha\right) \frac{e^{-\alpha r}}{r} = 0 \tag{8.26}$$

This equation can hold for all values of r only if each expression enclosed by parentheses vanishes, namely,

$$\alpha^2 + \frac{8\pi^2\mu E}{h^2} = 0 \qquad \frac{2\pi\mu e^2}{\epsilon_0 h^2} - 2\alpha = 0 \tag{8.27}$$

The second equation gives the value of α. We find that it turns out to be just the reciprocal of the first Bohr radius:

$$\alpha = \frac{\pi\mu e^2}{\epsilon_0 h^2} = \frac{1}{a_H} \tag{8.28}$$

Substituting this value of α into the first equation and solving for E, we obtain

$$E = -\frac{\mu e^4}{8\epsilon_0 h^2} \tag{8.29}$$

This is identical with the value of the energy of the first Bohr orbit, obtained in Section 8.3. It is the energy of the ground state of the hydrogen atom.

Now the solution given by Equation (8.25), in which α is given by Equation (8.28), does not yet represent a completely acceptable state function, for it is not normalized. But one can always multiply any solution by an arbitrary constant and still have a solution of the differential equation. Thus, by introducing a normalizing constant C, we can write

$$\psi = Ce^{-\alpha r} \tag{8.30}$$

This does not affect the value of the energy E. The normalizing condition equation (8.23) reduces to

$$\int_0^\infty C^2 e^{-2\alpha r} 4\pi r^2 \, dr = 1 \tag{8.31}$$

from which C may be found. If one is interested only in the spatial variation of the state function, it is not necessary to include the normalizing constant.

According to the above results the ground state of the hydrogen

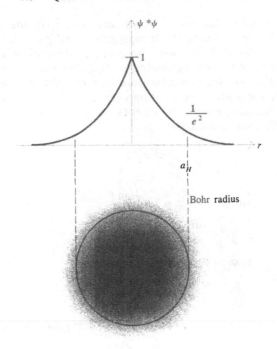

Figure 8.5. Probability density of the ground state (1s) of the hydrogen atom.

atom is such that the probability density of the electron is spherically symmetric and decreases exponentially with the radial distance r. The density is greatest at the center and diminishes by the factor e^{-2} in a distance of one Bohr radius. A plot of the density function is shown in Figure 8.5.

Excited States In order to find the state functions and the energies of the excited states of the hydrogen atom, it is necessary to solve the Schrödinger equation completely. To do this, the method of separation of variables is used.

The state function ψ is expressed as a product of three functions, a radial function and two angular functions, namely,

$$\psi(r,\theta,\phi) = R(r) \cdot \Theta(\theta) \cdot \Phi(\phi) \qquad (8.32)$$

The Schrödinger equation (8.23) can then be written as

$$\frac{1}{\Phi} \frac{d^2\Phi}{d\phi^2} + \frac{\sin\theta}{\Theta} \frac{d}{d\theta}\left(\sin\theta \frac{d\Theta}{d\theta}\right) + \frac{\sin^2\theta}{R} \frac{d}{dr}\left(r^2 \frac{dR}{dr}\right)$$
$$+ \sin^2\theta \frac{8\pi^2\mu r^2}{h^2}(E - V) = 0 \quad (8.33)$$

Now in order that our assumed solution satisfy the differential equation for all values of the independent variables r, θ, and ϕ, the first term $(1/\Phi)$ $(d^2\Phi/d\phi^2)$ must necessarily be equal to a constant. Otherwise r and θ would be dependent on ϕ. We denote this constant by $-m^2$. The remaining equation can then be split into two parts, a radial part and a part dependent only on θ. Setting each part equal to a second constant, denoted by $l(l + 1)$, one obtains the following separated differential equations:

$$\frac{1}{\Phi} \frac{d^2\Phi}{d\phi^2} = -m^2 \tag{8.34}$$

$$\frac{m^2}{\sin^2 \theta} - \frac{1}{\Theta \sin \theta} \frac{d}{d\theta} \left(\sin \theta \frac{d\Theta}{d\theta} \right) = l(l + 1) \tag{8.35}$$

$$\frac{1}{R} \frac{d}{dr} \left(r^2 \frac{dR}{dr} \right) + \frac{8\pi^2\mu r^2}{h^2} (E - V) = l(l + 1) \tag{8.36}$$

where, in the case of the hydrogen atom,

$$V = -\frac{e^2}{4\pi \epsilon_0 r}$$

A solution of the differential equation (8.34) involving ϕ is clearly

$$\Phi = e^{im\phi} \tag{8.37}$$

In order for this to be a physically acceptable solution, it is necessary that Φ assumes the same value for ϕ, $\phi + 2\pi$, $\phi + 4\pi$, and so forth; otherwise, the state function would not be uniquely defined at a given point in space. This requirement restricts the allowed values of m to integers, namely,

$$m = 0, \pm 1, \pm 2, \pm 3, \ldots \tag{8.38}$$

The number m is called the *magnetic quantum number*.

The θ-dependent equation (8.35) and the radial equation (8.36) are more difficult to solve. We shall not go into the details of their solution here, but shall merely give the results as given in any standard text on quantum mechanics. It happens that both differential equations were well known long before the time of Schrödinger. Their solutions had been worked out in connection with other problems in mathematical physics.

The equation in θ, Equation (8.35), is one form of an equation known as "Legendre's differential equation." This equation yields acceptable (single-valued) solutions only if l is a positive integer whose value is equal to or greater than $|m|$. The integer l is called the *azimuthal quantum number*. The resulting solutions are known as *associated Legendre polynomials*, denoted by $P_l^{|m|}$ (cos θ). They may be found by means of a generating formula

$$P_l^{|m|}(x) = \frac{(1 - x^2)^{(1/2)|m|}}{2^l l!} \left(\frac{d}{dx}\right)^{|m|+l} (x^2 - 1)^l \tag{8.39}$$

A few of these are:

$$P_0^0(x) = 1 \qquad\qquad P_1^1(x) = (1 - x^2)^{1/2}$$
$$P_1^0(x) = x \qquad\qquad P_2^1(x) = 3x(1 - x^2)^{1/2}$$
$$P_2^0(x) = (\tfrac{1}{2})(3x^2 - 1) \qquad P_2^2(x) = 3(1 - x^2)$$

The radial equation (8.36) is called "Laguerre's differential equation." Acceptable solutions (ones leading to quadratically integrable functions) are given by the formula

$$R(\rho) = \rho^l\, e^{-(1/2)\rho}\, L_{n+l}^{2l+1}(\rho)$$

The variable ρ is defined as a certain constant times r, namely,

$$\rho = \frac{2r}{na_H} \tag{8.40}$$

The quantity n is the *principal quantum number*. It is an integer whose value is equal to or greater than $l + 1$. The functions $L_{n+l}^{2l+1}(\rho)$ are called *associated Laguerre polynomials*. A formula for generating them is

$$L_{n+l}^{2l+1}(\rho) = \left(\frac{d}{d\rho}\right)^{2l+1} \left[e^\rho \left(\frac{d}{d\rho}\right)^{n+1} (\rho^{n+l}\, e^{-\rho})\right] \tag{8.41}$$

Some of these are as follows:

$$L_0^0(\rho) = 1 \qquad\qquad L_1^1(\rho) = -1$$
$$L_1^0(\rho) = 1 - \rho \qquad\quad L_2^1(\rho) = 2\rho - 4$$
$$L_2^0(\rho) = \rho^2 - 4\rho + 2 \quad L_2^2(\rho) = 2$$

In the process of solving the radial differential equation, it is found that the eigenvalues of the energy E are determined solely by the principal quantum number n. The eigenvalues are given by the formula

$$E_n = -\frac{\mu e^4}{8\epsilon_0 n^2}\left(\frac{1}{n^2}\right) \tag{8.42}$$

This is exactly the same formula as that given by the simple Bohr theory.

For each value of the principal quantum number n, with energy E_n given by the above equation, there are n different possible values of the azimuthal quantum number l, namely, $0, 1, 2, \ldots, n-2, n-1$. Each value of l represents a different kind of eigenstate. States in which $l = 0, 1, 2, 3$ are traditionally called s, p, d, and f states, respectively.

For each value of l, there are $2l + 1$ possible values of the magnetic quantum number m. These are $-l$, $-(l-1)$, ..., -1, $0, +1$, ..., $+(l-1)$, $+l$. As a result, there are n^2 different eigenfunctions or states for each value of n. The following diagram summarizes the situation:

n	1		2			3		
l	0 (s)	0 (s)	1 (p)	0 (s)	1 (p)		2 (d)	
m	0	0	-1 0 $+1$	0	-1 0 $+1$	-2 -1 0 $+1$ $+2$		

The complete state function, corresponding to prescribed values of n, l, and m, is given by the formula

$$\psi_{n,l,m} = C\rho^l \; e^{-\rho/2} \; L_{n+l}^{2l+1} (\rho) \; P_l^{|m|}(\cos \theta) \; e^{im\phi} \qquad (8.43)$$

Table 8.2. EIGENFUNCTIONS OF THE HYDROGEN ATOM (NORMALIZING CONSTANTS OMITTED)

—State	n	l	m		
1s	1	0	0	$e^{-\rho/2}$	
2s	2	0	0	$e^{-\rho/2} (1 - \rho)$	
2p	2	1	-1	$e^{-\rho/2} \rho$	$\begin{cases} \sin \theta \; e^{-i\phi} \\ \cos \theta \\ \sin \theta \; e^{+i\phi} \end{cases}$
			0		
			$+1$		
3s	3	0	0	$e^{-\rho/2} (\rho^2 - 4\rho + 2)$	
3p	3	1	-1	$e^{-\rho/2} (\rho^2 - 2\rho)$	$\begin{cases} \sin \theta \; e^{-i\phi} \\ \cos \theta \\ \sin \theta \; e^{+i\phi} \end{cases}$
			0		
			$+1$		
3d	3	2	-2	$e^{-\rho/2} \rho^2$	$\begin{cases} \sin^2 \theta \; e^{-i2\phi} \\ \sin \theta \cos \theta \; e^{-i\phi} \\ (1 - 3 \cos^2 \theta) \\ \sin \theta \cos \theta \; e^{+i\phi} \\ \sin^2 \theta \; e^{+i2\phi} \end{cases}$
			-1		
			0		
			$+1$		
			$+2$		

---

where $\rho = 2r/na_H$, and C is a normalizing constant. Table 8.2 shows a few of the simpler state functions of the hydrogen atom. Some of these are illustrated in Figure 8.6.

Angular Momentum   The angular momentum of an electron moving in a central force field can be obtained by standard quantum-

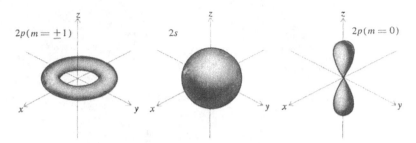

Figure 8.6. Probability density for the first excited state ($n = 2$) of the hydrogen atom.

mechanical methods. It develops that the angular momentum is quantized, as in the Bohr theory, and that the magnitude is determined by the quantum numbers designating the various quantum states.

The *total orbital angular momentum* is given by an expression involving only the azimuthal quantum number l, namely,

$$\sqrt{l(l+1)}\ \frac{h}{2\pi} = \sqrt{l(l+1)}\ \hbar$$

This is different from the value $nh/2\pi$ of the Bohr theory. In particular, for s states ($l = 0$), the angular momentum is zero. Physically this means that the electron cloud for s states does not possess a net rotation. It does not preclude any motion of the electron.

Theory also shows that the *z component of the angular momentum is quantized.* This component has the value

$$\frac{mh}{2\pi} = m\hbar$$

where m is the magnetic quantum number. As we have seen, m is the quantum number associated with the angle of rotation ϕ about the z axis. Since m can assume any of the values $0, \pm 1, \pm 2, \ldots, \pm l$, it follows that there are $2l + 1$ different possible values of the z component of the angular momentum for a given value of l. This is illustrated in Figure 8.7.

8.6   Radiative Transitions and Selection Rules

As already mentioned, when an atom is in the process of changing from one eigenstate to another, the probability density of the electronic charge becomes coherent and oscillates sinusoidally with a frequency given by the Bohr frequency condition. The way in which the charge cloud oscillates depends on the particular eigenstates involved. In the case of a so-called *dipole transition,* the centroid of the negative charge of the electron cloud oscillates about the pos-

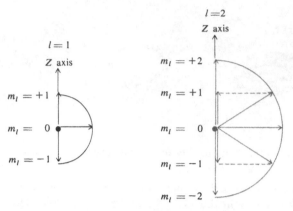

Figure 8.7. Space quantization of angular momentum for the cases $l = 1$ and $L = 2$.

itively charged nucleus. The atom thereby becomes an *oscillating electric dipole.*

Figure 8.8 is a diagram showing the time variation of the charge distribution for the hydrogen atom when it is in the coherent state represented by the combination $1s + 2p(m = 0)$. It is seen that the centroid of the charge moves back and forth along the z axis. The associated electromagnetic field has a directional distribution that is the same as that of a simple dipole antenna lying along the z axis. Thus the radiation is maximum in the xy plane and zero along the z axis. The radiation field in this case is linearly polarized with its plane of polarization parallel to the dipole axis.

A different case is shown in Figure 8.9. Here the coherent state is the combination $1s + 2p(m = +1)$. The centroid of the electronic charge now moves in a circular path around the z axis. The angular frequency of the motion is also that given by the Bohr frequency formula $\omega = \Delta E / \hbar$.

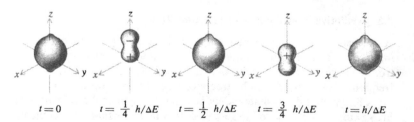

Figure 8.8. Charge distribution in the coherent state $1s + 2p_0$ as a function of time. The atom is an oscillating dipole.

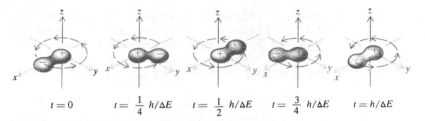

$$t = 0 \qquad t = \frac{1}{4} h/\Delta E \qquad t = \frac{1}{2} h/\Delta E \qquad t = \frac{3}{4} h/\Delta E \qquad t = h/\Delta E$$

Figure 8.9. Charge distribution in the coherent state $1s + 2p_1$ as a function of time. The atom is a rotating dipole.

Instead of an oscillating dipole, the atom is now a *rotating dipole*. The associated radiation field is such that the polarization is circular for radiation traveling in the direction of the z axis and linear for radiation traveling in a direction perpendicular to the z axis. For inter-

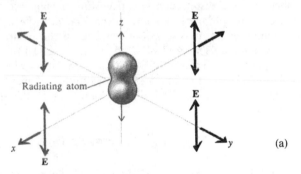

(a)

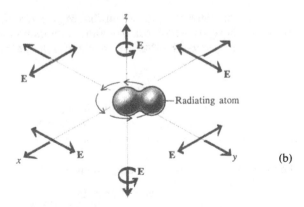

(b)

Figure 8.10. Polarization of the associated electromagnetic radiation (**E** vector) for (a) an oscillating dipole and (b) a rotating dipole.

mediate directions, the polarization is elliptical. The cases are illustrated in Figure 8.10. The coherent state $1s + 2p(m = -1)$ is just the same as the state $1s + 2p(m = +1)$ except that the direction of rotation of the electronic charge is reversed. Consequently, the sense of rotation of the associated circularly polarized radiation is also reversed.

In an ordinary spectral-light source the radiating atoms are randomly oriented in space and their vibrations are mutually incoherent. The total radiation is thus an incoherent mixture of all types of polarization. In other words the radiation is unpolarized. However, if the source is placed in a magnetic field, the field provides a preferred direction in space—the z axis in the above discussions [Figure 8.11(a)]. In addition the interaction between the radiating electron and the magnetic field causes each energy level to become split into several sublevels—one for each value of the magnetic quantum number m. As a result each spectrum line is split into several components. This splitting is known as the *Zeeman effect*. By means of the Zeeman effect, it is possible to observe the polarization effects mentioned above. This is illustrated in Figure 8.11(b).

The general theory of atomic emission and absorption involves the calculation of certain integrals known as *matrix elements*. The matrix element involved in electric dipole radiation is the quantity $\mathbf{M}_{AB}$ defined as

$$\mathbf{M}_{AB} = \iiint \psi_A{}^* \, e\mathbf{r}\psi_B \, dx \, dy \, dz \qquad (8.44)$$

where $\mathbf{r} = \hat{\mathbf{i}}x + \hat{\mathbf{j}}y + \hat{\mathbf{k}}z$, and e is the electric charge. The dipole matrix element is a measure of the amplitude of the oscillating dipole moment of the coherent state formed by the two stationary states ψ_A and ψ_B.

In the case of hydrogen it turns out that $\mathbf{M}_{AB}$ is zero for all pairs of states except those for which the azimuthal quantum numbers l_A and l_B differ by exactly one. In other words electric dipole transitions are "allowed" if

$$\Delta l = \pm 1 \qquad (8.45)$$

This is known as the *l-selection rule*. It implies that the angular momentum of the atom changes by an amount $\hbar$ during a dipole transition. This angular momentum is taken up by the photon involved in the transition.

There is also an *m*-selection rule, which is

$$\Delta m = 0 \text{ or } \pm 1 \qquad (8.46)$$

Transitions for which $\Delta m = 0$ are of the simple linear-dipole type,

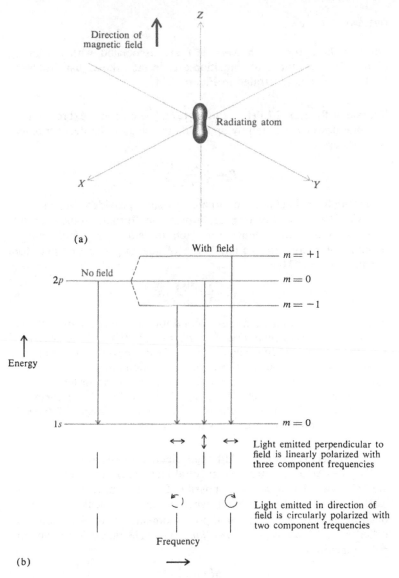

(a)

(b)

Figure 8.11. The Zeeman effect. Simplified diagram which neglects electron spin. (a) Direction of the external magnetic field; (b) transition diagram. For light that is emitted in a direction perpendicular to the magnetic field, there are three components that are linearly polarized as shown. There are only two components for the light emitted to the field, and these are circularly polarized as indicated. The fundamental Zeeman splitting is $\Delta\nu = eH/4\pi\mu_0 m$, where H is the magnetic field, e is the electronic charge, and m is the mass of the electron [40]. (Note: Effects of electron spin are omitted for reasons of simplicity.)

whereas those for which $\Delta m = \pm 1$ are associated with a rotating dipole. The two types of transition are, in fact, those just discussed for hydrogen and illustrated in Figure 8.11.

Transition Rates and Lifetimes of States   The classical expression for the total power P emitted by an oscillating electric dipole, of moment $M = M_0 \cos \omega t$, is

$$P = \frac{1}{3} \frac{\omega^4 M_0^2}{\pi \epsilon_0 c^3} \qquad (8.47)$$

The same formula applies to atomic emission, provided $|M_{AB}|$ is used for M_0. We must, however, interpret the formula somewhat differently in this case. Since for each transition an atom emits a quantum of energy $h\nu$, the number A of quanta per second per atom is equal to $P/h\nu$. Thus

$$A = \frac{2}{3} \frac{\omega^3 |M_{AB}|^2}{h\epsilon_0 c^3} \qquad (8.48)$$

is the number of transitions per second for each atom. This is known as the *transition probability*. The reciprocal, $1/A$, of the transition probability has the dimension of time. It is known as the *radiative lifetime* and is a measure of the time an excited atom takes to emit a light quantum. Typically, atomic lifetimes are of the order of 10^{-8} s for allowed dipole transitions in the visible region of the spectrum. The frequency factor ω^3 results in correspondingly longer lifetimes in the infrared and shorter lifetimes in the ultraviolet.

Higher-Order Transitions   Although electric dipole transitions generally give rise to the strongest spectral lines, such transitions are not the only ones that occur. It is possible for an atom to radiate or absorb electromagnetic radiation when it has an oscillating electric quadrupole moment, but no dipole moment. Such transitions are called *electric quadrupole transitions*. The selection rule for quadrupole transition is

$$\Delta l = \pm 2$$

It is easily shown that the charge distribution for a coherent state such as $1 s + 3d(m = 0)$ consists of an oscillating electric quadrupole. Transition probabilities for quadrupole radiation are usually several orders of magnitude smaller than those for electric dipole radiation. Lifetimes against quadrupole radiation are typically of the order of 1 s. Higher-order transitions such as octupole transitions, $\Delta l \pm 3$, and so forth, can also occur. Such transitions are seldom observed in connection with optical spectra, but they frequently occur in processes involving the atomic nucleus.

8.7 Fine Structure of Spectrum Lines. Electron Spin

If the spectrum of hydrogen is examined with an instrument of high resolving power, it is found that the lines are not single, but consist of several closely spaced components. The line Hα, for example, appears as two lines having a separation of about 0.14 Å. (This is not the same as the hydrogen–deuterium splitting discussed earlier.) This splitting of spectrum lines into several is known as *fine structure*. The lines of other elements besides hydrogen also possess a fine structure. These are designated as singlets, doublets, triplets, and so forth, depending on the number of components.

The theoretical explanation of fine structure was first made by Pauli, who postulated that the electron possesses an *intrinsic angular momentum* in addition to its orbital angular moment. This angular momentum is known as *spin*. All electrons have the same amount of spin, regardless of their motion, binding to atoms, and so forth. Theory shows that the component of this spin in a given direction must always be one or the other of the two values:

$$+ \tfrac{1}{2} \hbar \text{ or } - \tfrac{1}{2} \hbar$$

The total angular momentum of an electron in an atom then consists of the vector sum of its orbital angular momentum l and its spin s. The total angular momentum of a single electron, denoted by the symbol j, is then given by

$$\mathbf{j} = \mathbf{l} + \mathbf{s} \qquad (8.49)$$

It is customary to express the various angular momenta in units of $\hbar$. The angular momenta in these units are then essentially quantum numbers. For a given value of the azimuthal quantum number l, there are two values of the quantum number for the total angular momentum of a single electron, namely,

$$j = l + 1/2 \quad \text{and} \quad j = l - 1/2 \qquad (8.50)$$

Thus for $l = 1, j = 3/2$ or $1/2$, for $l = 2, j = 5/2$ or $3/2$, and so forth. For the case $l = 0$ there is only one value because the two states $j = +1/2$ and $j = -1/2$ are actually the same.

Now it was stated earlier that those states of hydrogen with a given value of the principal quantum number n, all have the same energy. This is not strictly true since the electron spin was not taken into account. Actually, as the electron undergoes its orbital motion around the positively charged nucleus, it experiences a magnetic field arising from this motion. The magnetic moment associated with the spin interacts with the magnetic field. This is called *spin-orbit* interaction.[2] The result of the spin-orbit interaction is that the two states $j = l + s$ and $j = l - s$ have slightly different energies. This in turn

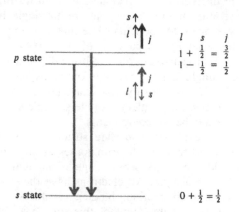

Figure 8.12. Fine structure (spin splitting) of a spectrum line for a $p \rightarrow s$ transition.

produces a splitting of the spectrum lines. A simplified diagram illustrating the splitting for the case of a $p \rightarrow s$ transition is shown in Figure 8.12.

8.8 Multiplicity in the Spectra of Many-Electron Atoms. Spectroscopic Notation

For atoms containing more than one electron, the total angular momentum $\mathbf{J}$ is given by the vector sum of all the individual orbital momenta $l_1, l_2, \ldots$, and spins $s_1, s_2, \ldots$, and so forth.

In the usual case the orbital angular momenta couple together to produce a resultant orbital angular momentum $\mathbf{L} = \mathbf{l_1} + \mathbf{l_2} + \ldots$. Similarly, the spins couple to form an overall resultant spin $\mathbf{S} = \mathbf{s_1} + \mathbf{s_2} + \ldots$. The total angular momentum is then given by the coupling of $\mathbf{L}$ and $\mathbf{S}$,

$$\mathbf{J} = \mathbf{L} + \mathbf{S} \tag{8.51}$$

This type of coupling is known as "LS coupling." Other types of coupling can also occur, such as "jj coupling" in which the individual js

---

[2] In hydrogen, in addition to the spin-orbit splitting of the energy levels there is also a splitting due to relativistic effects. This relativity splitting is also very small and causes the energies of states with the same n but different l to be slightly different [24].

add together to produce a resultant **J**. In general, *LS* coupling occurs in the lighter elements and *jj* coupling in the heavy elements.

In *LS* coupling all three quantities **L**, **S**, and **J** are quantized. Their magnitudes are given by

$$|\mathbf{L}| = \hbar\sqrt{L(L+1)}, \ |\mathbf{S}| = \hbar\sqrt{S(S+1)}, \ |\mathbf{J}| = \hbar\sqrt{J(J+1)}$$

where L, S, and J are quantum numbers with the following properties.

The quantum number L is always a positive integer, or zero. The spin quantum number S is either integral or half integral, depending on whether the number of electrons is even or odd, respectively. Consequently, the total angular-momentum quantum number J is integral, or half integral, depending on whether there is an even or odd number of electrons, respectively.

The total energy of a given state depends on the way the various angular momenta add together to produce the resultant total angular momentum. Hence, for given values of L and S the various values of J correspond to different energies. This in turn results in the fine structure of the spectral lines.

The spectroscopic designation of a state having given values of L, S, and J is the following:

$$^{2S+1}L_J$$

Here the quantity $2S + 1$ is known as the *multiplicity*. It is the number of different values that J can assume for a given value of L, provided $L \gg S$, namely,

$$L + S, L + S - 1, L + S - 2, \ldots L - S$$

If $L < S$, then there are only $2L + 1$ different J values, namely,

$$L + S, L + S - 1, L + S - 2, \ldots |L - S|$$

This is known as "incomplete multiplicity."

If $S = 0$, the multiplicity is unity. The state is then said to be a "singlet." Similarly, for $S = \frac{1}{2}$, the multiplicity is two, and the state is a doublet. Table 8.3 lists the spin, multiplicity, and names of the first few types of states.

For one-electron atoms, only one value of S is possible, namely, $\frac{1}{2}$. Hence all states of one-electron atoms are doublet states. In the case of two electrons, S can have either of the two values $\frac{1}{2} + \frac{1}{2} = 1$ or $\frac{1}{2} - \frac{1}{2} = 0$. Thus for two-electron atoms there are two sets of states, triplets and singlets.

In addition to the naming of states according to multiplicity, a letter is used to designate the value of the total orbital angular momentum L. This designation is given in Table 8.4.

The letter S, designating states for $L = 0$, is not the spin quantum

Table 8.3. MULTIPLICITIES OF STATES

S	Multiplicity $(2S + 1)$	Name
0	1	Singlet
$\frac{1}{2}$	2	Doublet
1	3	Triplet
$\frac{3}{2}$	4	Quartet
2	5	Quintet
$\frac{5}{2}$	6	Sextet

number, although it is the same letter. This is confusing, but it is accepted convention.

Let us consider, as an example, the case of two electrons. Let one electron be a p electron ($l_1 = 1$) and the other be a d electron ($l_2 = 2$). The possible values of L are $l_1 + l_2$, $|l_1 - l_2|$, and all integral values between. Thus $L = 1$, 2, or 3, which means that we have P states, D states, and F states. Since S can be 0 or 1, then there are both singlets and triplets for each L value. The complete list of possible states for the combination of a p and a d electron is the following:

Singlets	Triplets		
$^1P_1$	$^3P_0$	$^3P_1$	$^3P_2$
$^1D_2$	$^3D_1$	$^3D_2$	$^3D_3$
$^1F_3$	$^3F_2$	$^3F_3$	$^3F_4$

Selection Rules In the case of LS coupling, the selection rules that govern allowed transition for dipole radiation are the following:

$$\Delta L = 0, \pm 1$$
$$\Delta S = 0$$
$$\Delta J = 0, \pm 1 \qquad (J = 0 \rightarrow J = 0 \text{ forbidden})$$

Table 8.4. DESIGNATION OF STATES ACCORDING TO ORBITAL ANGULAR MOMENTUM

L:		0	1	2	3	4	5	6	7	8	·	·
Designation		S	P	D	F	G	H	I	K	M	·	·

In all cases the symbol Δ means the difference between the corresponding quantum numbers of the initial and final states of the transition.

Parity In addition to the above selection rules there is another important rule involving a concept known as *parity*. The parity of an atomic state can be even or odd. This is determined by the sum of the l values of the individual electrons. If the sum is even (odd), the parity is even (odd). For example, consider the states of a two-electron atom. If one electron is an s electron ($l_1 = 0$) and the other is a p electron ($l_2 = 1$), then $l_1 + l_2 = 1$, hence all sp states are of odd parity. Similarly, all sd states are of even parity, and so forth. The following selection rule holds for electric dipole radiation from transitions between two states:

$$\text{even} \leftrightarrows \text{odd (allowed)} \qquad \left.\begin{array}{l} \text{odd} \rightarrow \text{odd} \\ \\ \text{even} \rightarrow \text{even} \end{array}\right\} \quad \text{(forbidden)}$$

In other words the parity of the final state must be different from the parity of the initial state.

It is possible for an excited state to be such that it cannot undergo a transition by dipole radiation to any lower state. In this case the state is said to be *metastable*. If an atom is in a metastable state, it must return to the ground state either by emission of radiation other than dipole radiation — for example, quadrupole, and so forth — or it may return via collisions with other atoms.

8.9 Molecular Spectra

Molecules, like atoms, are found to exhibit discrete frequency spectra when appropriately excited in the vapor state. This indicates that the energy states of molecules are quantized and that a molecule may emit or absorb a photon upon changing from one energy state to another.

For purposes of spectroscopy, the energy of a molecule may be expressed as the sum of three kinds of energy. These are rotational energy E_{rot}, vibrational energy E_{vib}, and electronic energy E_{el}. Thus

$$E = E_{\text{rot}} + E_{\text{vib}} + E_{\text{el}}$$

Of the three, E_{rot} is generally the smallest, typically a few hundredths of an electron volt. Vibrational energies are of the order of tenths of an electron volt, whereas the largest energies are of the electronic type that are generally a few electron volts. Each of the three types of energy is quantized in a different way and, accordingly, each is associated with a different set of quantum numbers.

It is possible for transitions to take place that involve the rotational energy levels only. The resulting spectrum, called the *pure rotation spectrum,* usually lies in the far infrared or the microwave region. If vibrational changes also occur in the transition but no electronic changes, we have the *rotation–vibration spectrum.* The rotation–vibration lines are generally found in the near infrared. Finally, transitions that involve electronic-energy changes are the most energetic. Lines of the *electronic spectra* of molecules occur typically in the visible and ultraviolet regions.

Rotational Energy Levels   The rotational energy, E_{rot}, is the kinetic energy of rotation of the molecule as a whole. The quantization of the rotational energy is expressed in terms of rotational quantum numbers. How many of these rotational quantum numbers are needed to specify a given rotational state depends on the particular molecular geometry. There are four basic molecular types. These are

(1) Linear molecules
(2) Spherical-top molecules
(3) Symmetrical-top molecules
(4) Asymmetrical-top molecules

The four types are illustrated in Figure 8.13.

In the case of linear molecules and spherical-top molecules, only one quantum number J is needed to specify the rotational state. As

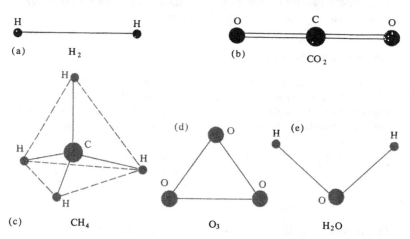

Figure 8.13.   Some cases to illustrate the different types of molecular rotational symmetry. (a), (b) Linear molecules: $I_a = I_b$, $I_c = 0$; (c) spherical top: $I_a = I_b = I_c$; (d) symmetrical top: $I_a = I_b \neq I_c$; (e) asymetrical top: $I_a \neq I_b \neq I_c$.

with atomic states, the quantity $\hbar\sqrt{J(J+1)}$ is the magnitude of the rotational angular momentum. The rotational energy is given by the quantum equivalent of the classical value, namely,

$$E_{\text{rot}} = \frac{(\frac{1}{2})\left[\hbar\sqrt{J(J+1)}\right]^2}{I} = J(J+1)\,Bhc \qquad (8.52)$$

where

$$B = \frac{h}{8\pi^2 cI} \qquad (8.53)$$

and

$$J = 0, 1, 2, \ldots$$

Here I is the moment of inertia of the molecule about the axis of rotation. (For a symmetrical diatomic molecule consisting of two atoms of mass $M/2$ separated by a distance $2b$, the moment of inertia is given by the classical expression $I = Mb^2$.) An energy-level diagram showing the rotational levels of a linear molecule is shown in Figure 8.14.

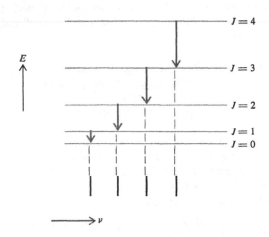

Figure 8.14.   Transition diagram for a pure rotational spectrum.

In the case of symmetrical-top molecules, two quantum numbers are needed to specify the rotational states. These are customarily written as J and K, where, again, $\hbar\sqrt{J(J+1)}$ is the total rotational angular momentum. The quantum number K is the component in units of $\hbar$ of the rotational angular momentum about the symmetry axis. For a given value of K, J can assume any of the values K, $K+1$, $K+2$, and so forth. The rotational energy levels are then given by the formula

$$E_{rot} = J(J + 1) Bhc + K^2(C - B)hc \qquad \text{(8.54)}$$

In the above formula the quantities B and C are related to the two principal moments of inertia of the molecule,

$$B = \frac{h}{8\pi^2 c I_b} \qquad C = \frac{h}{8\pi^2 c I_c} \qquad \text{(8.55)}$$

Here I_c is the moment of inertia about the symmetry axis, and I_b is the moment about the perpendicular axis.

In the case of the asymmetrical-top molecule, there are three different moments of inertia, and three rotational quantum numbers are involved. The theory is quite complicated in this case, and there is no simple formula for the energies of the quantized states [23].

Rotational transition rules are governed by the general selection rules

$$\begin{aligned} \Delta J &= 0, \pm 1 \\ \Delta K &= \pm 1 \end{aligned} \qquad \text{(8.56)}$$

In addition to the above rules, there are other selection rules involving the symmetry of the rotational states. We shall not go into a discussion of these.

Vibrational Energy Levels If a molecule contains N atoms, there are $3N$ modes of motion. Of these, three correspond to translation of the molecule, and three to rotation (or two for a linear molecule). The remaining $3N - 6$ (or $3N - 5$) correspond to the normal vibrational modes.

Theory shows that the quantizing of each vibrational mode can be expressed in terms of a single associated quantum number. The normal frequencies are designated $\nu_1, \nu_2, \ldots$, and so forth, and the associated vibrational quantum numbers are $v_1, v_2, \ldots$, and so on. The vibrational energy is then given by

$$E_{vib} = (v_1 + \tfrac{1}{2})h\nu_1 + (v_2 + \tfrac{1}{2})h\nu_2 + \cdots \qquad \text{(8.57)}$$

The above formula is valid if the vibrational displacements are small enough so that the motion is essentially harmonic in character. It indicates that the energy levels associated with a given normal mode are (1) equally spaced, and (2) the energy of the lowest vibrational state ($v_1 = 0$, $v_2 = 0$, . . .) is *not zero,* but has a finite value: $(\tfrac{1}{2})h\nu_1 + (\tfrac{1}{2})h\nu_2 + \ldots$. This energy is called the *zero-point energy.* It is present even at the absolute zero of temperature.

The selection rule for vibration transitions is

$$\Delta v = \pm 1 \qquad \text{(8.58)}$$

This rule is strictly adhered to only if the motion is absolutely harmonic. Such is never actually the case. Transitions in which $\Delta v = \pm 2, \pm 3$, and so forth, also occur. These *overtone transitions* are generally much weaker than the fundamental transitions in which $\Delta v = \pm 1$.

A homonuclear diatomic molecule does not exhibit a pure rotation spectrum nor does it show a rotation–vibration spectrum. This is because homonuclear molecules have no permanent electric dipole moment. Consequently neither rotational nor vibrational transitions produce an oscillating dipole moment. Thus there is no associated dipole radiation.

On the other hand heteronuclear diatomic molecules such as hydrogen chloride do exhibit strong rotation–vibration spectra.

A transition diagram illustrating the vibrational energy levels of a diatomic molecule, with rotational energy levels superimposed, is shown in Figure 8.15. The J-selection rule for rotation–vibration transitions is

$$\Delta J = 0, \pm 1 \qquad (8.59)$$

The spectrum is divided into three branches known as the P, Q, and R branches. They are determined by the value of ΔJ as follows:

$$\Delta J = -1 \qquad P \text{ branch}$$
$$\Delta J = 0 \qquad Q \text{ branch}$$
$$\Delta J = +1 \qquad R \text{ branch}$$

Electronic Energy States in Molecules The following discussion is restricted largely to the case of diatomic molecules, although many of the general principles apply to other molecules as well.

In molecules, the orbital angular momenta and spins of the electrons couple together in much the same manner that they do in atoms. In diatomic molecules, an important quantum number is one designating the sum of the projections of the orbital angular momenta on the line connecting the two atoms. This quantum number is denoted by the symbol Λ. The various electronic states corresponding to different values of Λ are designated by capital Greek letters as follows:

$$\Lambda : 0, 1, 2, 3, \ldots$$
Electronic state $: \Sigma, \Pi, \Delta, \Phi, \ldots$

For a given value of Λ, the rotational quantum number J can have any of the values $\Lambda, \Lambda + 1, \Lambda + 2$, and so forth.

As in the case of atoms, the total spin S determines the multiplicity of an electronic state. This multiplicity is $2S + 1$. It is the number of sublevels for a given value of J. Thus there are singlet

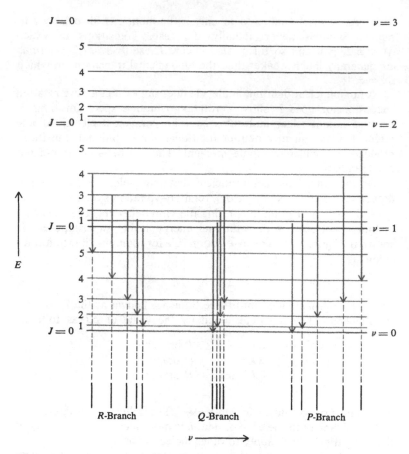

Figure 8.15. Transitional diagram for a rotation–vibration spectrum.

states ($S = 0$):

$$^1\Sigma, \ ^1\Pi, \ ^1\Delta, \ . \ . \ .$$

doublet states ($S = 1/2$):

$$^2\Sigma, \ ^2\Pi, \ ^2\Delta, \ . \ . \ .$$

and so on. It is also true, as with atoms, that the multiplicity is always odd (even) if the total number of electrons is even (odd).

The selection rules governing electronic transitions are

$$\Delta\Lambda = 0, \pm 1 \tag{8.60}$$
$$\Delta S = 0 \tag{8.61}$$

The following are examples of allowed electronic transitions:

$$^1\Sigma \rightarrow {}^1\Pi,\ {}^2\Pi \rightarrow {}^2\Delta,\ {}^3\Pi \rightarrow {}^3\Sigma$$

Since the electronic energies in molecules have the rotational and vibrational energies superimposed on them, electronic transitions may be accompanied by vibrational and rotational transitions. This results in a very large number of lines for each electronic transition and gives rise to the vibration–rotation structure in the electronic spectra of molecules. A partial energy-level diagram of the nitrogen molecule N_2 is shown in Figure 8.16 as an example.

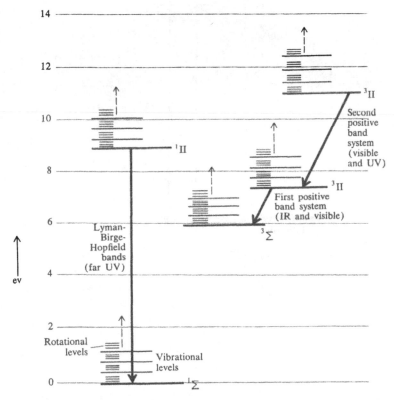

Figure 8.16. Partial energy-level diagram of the nitrogen molecule N_2. Electronic transitions of some of the important band systems are indicated. The rotational and vibrational energy levels are not drawn to scale.

8.10 Atomic Energy Levels in Solids

Consider an atom that is embedded in a solid, either as a part of the structure or as an impurity. One or more of the electrons may be shared by the solid as a whole, and thus are not associated with any particular atom.

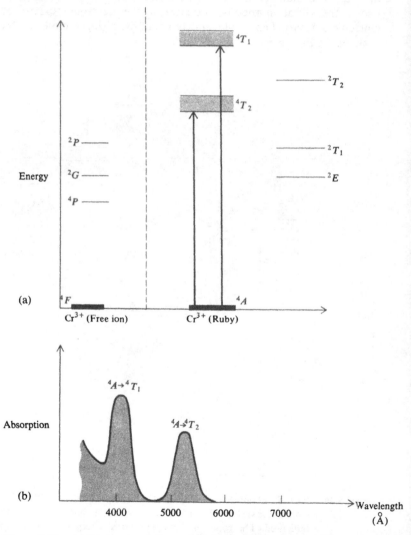

Figure 8.17. (a) Energy level diagram of the Cr^{3+} ion in the free state and in ruby; (b) absorption of the Cr^{3+} ion in ruby.

The energy levels of these electrons become smeared out into bands – the valence and conduction bands – of the solid. The atom in question then becomes an ion. The bound electrons associated with the ion may have various quantized states available to them and therefore various energy levels. This gives rise to a characteristic ionic absorption spectrum.

In the case of the rare earth ions, the unfilled shells involve deep-lying $4f$ electrons. These electrons are well shielded by the outer electrons, and the energy levels of the free ion are essentially unchanged when the ion is embedded in a solid. The levels are quantized according to angular momentum and are designated in the same way as the regular atomic energy levels discussed in Section 8.2.

For the transition metals, such as iron, chromium, and so forth, the $3d$ shell is unfilled. This shell is not as well shielded as the $4f$ shell in the rare earths. The result is that the energy levels of transition metal ions are profoundly changed when the ion is placed in a solid. Rather than angular momentum, it is the *symmetry* of the wave function that is important in the determination of the energy levels in this case, particularly if the ion is in a crystal lattice. The quantization of the energy is then largely dictated by the symmetry of the field due to the surrounding ions.

It is beyond the scope of this book to develop the theory of atomic energy levels in crystals. The subject is very involved. An extensive, rapidly growing literature already exists. However, as an illustration of the energy-level scheme of a typical case, the levels of the Cr^{3+} ion are shown in Figure 8.17. At the left are shown the levels of the free ion, while at the right are the levels of the Cr^{3+} ion in ruby. The symbols A, E, and T refer to different types of symmetry. The ruby crystal consists of Al_2O_3 (corundum) in which part of the aluminum atoms have been replaced with chromium. The red color of ruby is due to the absorption of green and blue light corresponding to transitions from the ground state, ^4A, to the excited states $^4T_1$ and $^4T_2$, as indicated.

For further reading in atomic, molecular, and solid-state spectroscopy, References [18] [19] [23] [24] [30] and [40] are recommended.

PROBLEMS

8.1 If R is the Rydberg constant for a nucleus of infinite mass, Equation (8.9), show that the Rydberg constant for a nucleus of mass M is given approximately by $R_M \approx R - (m/M)R$.

8.2 Calculate the difference between the frequencies of the Balmer $- \alpha$ lines of hydrogen and deuterium.

8.3 Calculate the frequency of the hydrogen transition $n = 101 \to n = 100$.

8.4 Show that the energy of the $2s$ state of atomic hydrogen (Table 8.2) is $-\frac{1}{4} R$ by substitution in the radial Schrödinger equation (8.36).

8.5 Consider a hydrogen atom in the ground state and imagine a sphere of radius r centered at the nucleus. Derive a formula for the probability that the electron is located inside the sphere. (a) What is the probability for $r = a_H$? (b) For what value of r is the probability equal to 99 percent?

8.6 Determine all of the states of a pf configuration of two electrons in LS coupling.

8.7 Find all of the allowed dipole transitions between a pd and a pf configuration.

8.8 Calculate the lifetime of the $2p$ state of atomic hydrogen by assuming that the magnitude of the dipole moment of the transition to the $1s$ state is approximately equal to ea_H.

8.9 Find the frequency of the radiation emitted by the pure rotational transition $J = 1 \to J = 0$ in hydrogen chloride. The distance between the hydrogen atom and the chloride atom is 1.3 Å.

CHAPTER 9

Amplification of Light. Lasers

9.1 Introduction

Few developments have produced an impact on any established field of science compared to the effect of the laser or optical maser[1] on the field of optics. Vacuum-tube oscillators that generate coherent electromagnetic radiation at frequencies up to about 10^9 Hz have been known for many years. In 1954 the maser was developed [13]. The maser generates microwaves (10^9 to 10^{11} Hz). The practical feasibility using the maser principle for the amplification of light ($\sim 10^{14}$ Hz) was studied in 1958 by Schawlow and Townes who laid down the basic theoretical groundwork. The first working laser, made of synthetic ruby crystal, was produced in 1960 at the Hughes Research Laboratories. It was followed in a few months by the helium–neon gas laser developed at the Bell Telephone Laboratories. The ruby laser generates visible red light. The helium–neon laser generates both visible red light and infrared radiation. Since their introduction, numerous types of lasers have been produced that generate radiation at various optical frequencies extending from the far infrared to the ultraviolet region of the spectrum [25].

A laser is basically an optical oscillator. It consists essentially of an amplifying medium placed inside a suitable optical resonator or cavity. The medium is made to amplify by means of some kind of external excitation. The laser oscillation can be described as a standing wave in the cavity. The output consists of an intense beam of highly monochromatic radiation.

Conventional light sources (arcs, filaments, discharges) provide luminous intensities corresponding to thermal radiation at temperatures of no more than about 10^4 °K. With lasers, intensities corresponding to 10^{20} to 10^{30} degrees are readily attained. Such enormous

---

[1] The word *laser* is an acronym for "light amplification by stimulated emission of radiation." Lasers were developed several years after *masers* (microwave amplifiers). For this reason the first lasers were called *optical masers*, but the single word *laser* is now the generally accepted usage.

values make possible the investigation of new optical phenomena such as nonlinear optical effects, optical beating, long-distance interference, and many others that were previously considered out of the question. Practical applications of the laser include long-distance communications, optical radar, microwelding, and eye surgery, to mention only a few.

9.2   Stimulated Emission and Thermal Radiation

Einstein in 1917 first introduced the concept of stimulated or induced emission of radiation by atomic systems. He showed that in order to describe completely the interaction of matter and radiation, it is necessary to include that process in which an excited atom may be induced, by the presence of radiation, to emit a photon, and thereby decay to a lower energy state.

Consider a quantized atomic system in which there are levels labeled 1,2,3, . . . , with energies E_1, E_2, E_3, and so forth. The populations, that is, the number of atoms per unit volume, in the various levels, are N_1, N_2, N_3, and so forth. If the atomic system is in equilibrium with thermal radiation at a given temperature T, then the relative populations of any two levels, say 1 and 2, are given by Boltzmann's equation

$$\frac{N_2}{N_1} = \frac{e^{-E_2/kT}}{e^{-E_1/kT}} \tag{9.1}$$

where k is Boltzmann's constant. If we assume, for definiteness, that $E_2 > E_1$, then $N_2 < N_1$.

An atom in level 2 can decay to level 1 by emission of a photon. Let us call A_{21} the transition probability per unit time for spontaneous emission from level 2 to level 1. Then the number of *spontaneous* decays per second is $N_2 A_{21}$. [The value of A_{21} can be calculated from Equation (8.8) of the previous chapter.]

In addition to these spontaneous transitions, there will be *induced* or *stimulated* transitions. The total rate of these induced transitions between level 2 and level 1 is proportional to the energy density u_ν of the radiation of frequency ν, where

$$\nu = \frac{(E_2 - E_1)}{h} \tag{9.2}$$

Let B_{21} and B_{12} denote the proportionality constants for stimulated emission. Then the number of stimulated downward transitions (emissions) per second is

$$N_2 B_{21} \, u_\nu$$

Similarly, the number of stimulated upward transitions (absorptions) per second is

$$N_1 B_{12} \, u_\nu$$

The proportionality constants in the above expressions are known as the Einstein A and B coefficients.

Under equilibrium conditions the net rate of downward transitions must be equal to that of upward transitions, namely,

$$N_2 A_{21} + N_2 B_{21} u_\nu = N_1 B_{12} u_\nu \qquad (9.3)$$

By solving for u_ν, we obtain

$$u_\nu = \frac{N_2 A_{21}}{N_1 B_{12} - N_2 B_{21}}$$

Further, in view of Equation (9.1), we can write

$$u_\nu = \frac{A_{21}}{B_{21}} \frac{1}{(B_{12}/B_{21}) \, e^{h\nu/kT} - 1} \qquad (9.4)$$

In order for this to agree with the Planck radiation formula, the following equations must hold:

$$B_{12} = B_{21} \qquad (9.5)$$

$$\frac{A_{21}}{B_{21}} = \frac{8\pi h\nu^3}{c^3} \qquad (9.6)$$

Thus for atoms in equilibrium with thermal radiation, the ratio of stimulated emission rate to spontaneous emission rate is given by the formula

$$\frac{\text{stimulated emission}}{\text{spontaneous emission}} = \frac{B_{21} u_\nu}{A_{21}} = \frac{1}{e^{h\nu/kT} - 1} \qquad (9.7)$$

We recall from Section 7.5 that this is precisely the same as the number of photons per mode, that is, the occupation index.

According to the above result, the rate of induced emission is extremely small in the visible region of the spectrum with ordinary optical sources ($T \sim 10^3 \, °K$). Hence, in such sources most of the radiation is emitted through spontaneous transitions. Since these transitions occur in a random manner, ordinary sources of visible radiation are incoherent.

On the other hand, in a laser the radiation density for certain preferred modes builds up to such a large value that induced transitions become completely dominant. One result is that the emitted radiation is highly coherent. Another is that the spectral radiance at the operating frequency of the laser is vastly greater than that of ordinary light sources.

9.3 Amplification in a Medium

Consider an optical medium through which radiation is passing. Suppose that the medium contains atoms in various energy levels, E_1, E_2, E_3, and so forth. Let us fix our attention on two levels, say E_1 and E_2, where $E_1 < E_2$. We have already seen that the rates of stimulated emission and absorption involving these two levels are proportional to $N_2 B_{21}$ and $N_1 B_{12}$, respectively. Since $B_{21} = B_{12}$, the rate of stimulated downward transitions will exceed that of upward transitions if

$$N_2 > N_1$$

that is, if the population of the upper state is greater than that of the lower state.[2]

Such a condition is contrary to the thermal equilibrium distribution given by the Boltzmann equation (9.1). It is called a *population inversion* (Figure 9.1). If a population inversion exists, then, as we shall

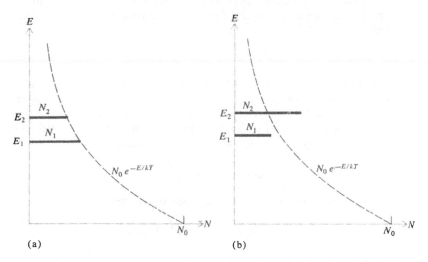

(a)   (b)

Figure 9.1. Graphs of the population densities of two levels of a system. (a) Normal or Boltzmann distribution; (b) inverted distribution.

show, a light beam will increase in intensity or, in other words, it will be amplified as it passes through the medium. This is because the gain due to the induced emission exceeds the loss due to absorption.

[2] If the energy levels of the system are *degenerate*, that is, if there are several sublevels belonging to a given energy, then the gain formulas must be modified as follows: The quantity N_1 is to be replaced by $(g_2/g_1) N_1$; the degeneracy parameters g_1 and g_2 are the numbers of sublevels in level 1 and level 2, respectively; the condition for amplification then becomes $N_2 > (g_2/g_1) N_1$.

The induced radiation is emitted in the same direction as the primary beam and it also has a definite phase relationship; that is, it is coherent with the primary radiation. This can be argued on the basis of the modes of the electromagnetic radiation field. *The act of stimulated emission of a single atom results in the addition of a photon to the particular mode that causes the stimulated emission.* As we have shown, the rate of stimulated emission is proportional to the number of photons in the mode in question. Modes are distinguished from one another by frequency, direction of the wave vector, and polarization. Hence a photon that has been added to a given mode by stimulated emission is a copy of the photons that are already in that mode.

The Gain Constant   In order to determine quantitatively the amount of amplification in a medium, we must take a closer look at the details of emission and absorption. Suppose a parallel beam of light propagates through a medium in which there is a population inversion. For a collimated beam, the spectral energy density u_ν is related to the spectral irradiance I_ν in the frequency interval ν to $\nu + \Delta\nu$ by the formula

$$u_\nu \, \Delta\nu = \frac{I_\nu \, \Delta\nu}{c} \tag{9.8}$$

Due to the Doppler effect and other line-broadening effects, not all of the atoms in a given energy level are effective for emission or absorption in a specified frequency interval. Rather, a certain number per unit volume, say ΔN_1, of the N_1 atoms per unit volume in level 1, is available. Consequently, the rate of upward transitions is

$$B_{12} u_\nu \, \Delta N_1 = B_{12} \left(\frac{I_\nu}{c}\right) \Delta N_1$$

Similarly, the rate of induced downward transitions is

$$B_{21} u_\nu \, \Delta N_2 = B_{21} \left(\frac{I_\nu}{c}\right) \Delta N_2$$

Now each upward transition subtracts a quantum of energy $h\nu$ from the beam. Similarly, each downward transition adds the same amount. Therefore the net time rate of change of the spectral energy density in the interval $\Delta\nu$ is given by

$$\frac{d}{dt} (u_\nu \, \Delta\nu) = h\nu (B_{21} \, \Delta N_2 - B_{12} \, \Delta N_1) u_\nu \tag{9.9}$$

In time dt the wave travels a distance $dx = c \, dt$. Hence, in view of Equation (9.8) we can write

$$\frac{dI_\nu}{dx} = \frac{h\nu}{c} \left(\frac{\Delta N_2}{\Delta\nu} - \frac{\Delta N_1}{\Delta\nu}\right) B_{21} I_\nu \tag{9.10}$$

giving the rate of growth of the beam in the direction of propagation.

The above differential equation can be integrated to give

$$I_\nu = I_{0\nu}\, e^{\alpha_\nu\, x} \tag{9.11}$$

in which α_ν is the gain constant at frequency ν. It is given by

$$\alpha_\nu = \frac{h\nu}{c}\left(\frac{\Delta N_2}{\Delta \nu} - \frac{\Delta N_1}{\Delta \nu}\right) B_{12} \tag{9.12}$$

An approximate expression for the gain constant at the center of a spectral line is obtained by taking $\Delta\nu$ to be the linewidth. The ΔNs are then set equal to the Ns. The result, which is correct except for a numerical constant of the order of unity, is

$$\alpha_{\max} \approx \frac{h\nu}{c\Delta\nu}(N_2 - N_1)B_{12} = \frac{\lambda^2}{8\pi\Delta\nu}(N_2 - N_1)A_{12} \tag{9.13}$$

The last step follows from the relation between the Einstein A and B coefficients, Equation (9.6).

We see that α is positive if $N_2 > N_1$, which is the condition for amplification. Otherwise, if $N_2 < N_1$ (which is the normal equilibrium condition), then α is negative and we have absorption. Methods for producing population inversions in optical media will be discussed in the following section.

The Gain Curve   In order to determine how the gain varies with frequency, it is necessary to consider the details of line broadening. In the case of broadening due to thermal motion alone, elementary kinetic theory [31] gives the fraction of atoms whose x component of velocity lies between u_x and $u_x + \Delta u_x$. This fraction is a Gaussian function, namely,

$$C e^{-a u_x^2}\, \Delta u_x$$

in which $C = (m/2\pi kT)^{1/2}$ and $a = m/2kT$. Here T is the absolute temperature and k is Boltzmann's constant. Due to the Doppler effect, these atoms will emit or absorb radiation, propagating in the x direction, of slightly different frequency ν than the resonance frequency ν_0 of the atom when it is at rest. The frequency difference is given by

$$\frac{\nu - \nu_0}{\nu_0} = \frac{u_x}{c}$$

It follows that the number of atoms in a given level that can absorb or emit in the frequency range ν to $\nu + \Delta\nu$ is given by

$$\Delta N_i = N_i C e^{-\beta(\nu-\nu_0)^2}\, \frac{c}{\nu_0}\, \Delta\nu$$

in which $\beta = mc^2/(2kT\nu_0{}^2)$. This can be substituted for ΔN_1 and ΔN_2 in Equation (9.12). The result is

$$\alpha_\nu = Ce^{-\beta(\nu-\nu_0)^2}(N_2 - N_1)hB_{21} \qquad (9.14)$$

Thus the gain for a Doppler-broadened laser transition varies with frequency according to a Gaussian function. A curve is shown in Figure 9.2. This same curve also represents the profile of a Doppler-

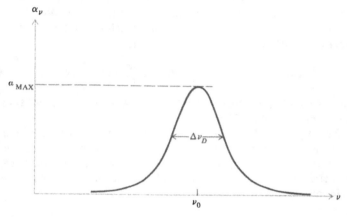

Figure 9.2. Amplification coefficient for a Doppler broadened spectral line.

broadened spectral line. The maximum gain occurs at the line center and is given by

$$\alpha_{max} = C(N_2 - N_1)hB_{21} = C(N_2 - N_1)\frac{\lambda_0{}^3}{8\pi}A_{21} \qquad (9.15)$$

In order to compare this with the approximate value expressed by Equation (9.13) we must calculate the width of a Doppler-broadened line. The width at half maximum is found by setting the exponential factor $e^{-\beta(\nu-\nu_0)^2}$ equal to $\frac{1}{2}$, that is, $\nu - \nu_0 = (\ln 2/\beta)^{1/2}$. Twice this value is the full width $\Delta\nu_D = 2(\ln 2/\beta)^{1/2}$. On using the definitions of C and β given above, we readily find

$$\alpha_{max} = 2\sqrt{\frac{\ln 2}{\pi}}\frac{\lambda_0{}^2}{8\pi\Delta\nu_D}(N_2 - N_1)A_{21} \qquad (9.16)$$

The numerical factor $2(\ln 2/\pi)^{1/2} = 0.939$, hence the previous formula (9.13) is quite accurate in this case.

9.4 Methods of Producing a Population Inversion

There are several methods for producing the population inversions necessary for optical amplification to take place. Some of the most commonly used are

(1) Optical pumping or photon excitation
(2) Electron excitation
(3) Inelastic atom–atom collisions
(4) Chemical reactions

In the case of optical pumping, an external light source is employed to produce a high population of some particular energy level in the laser medium by selective optical absorption [Figure 9.3(a)]. This is the

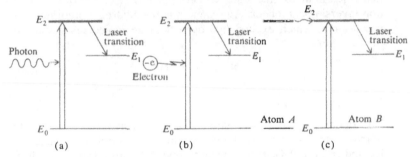

(a)   (b)   (c)

Figure 9.3. Diagrams showing three processes for producing a population inversion. (a) Optical pumping; (b) direct electron excitation; (c) inelastic atom–atom collisions.

method of excitation used in the solid-state lasers of which the ruby laser is the prototype.

Direct electron excitation in a gaseous discharge may be used to produce the desired inversion [Figure 9.3(b)]. This method is used in some of the gaseous ion lasers such as the argon laser. With this type of excitation the laser medium itself carries the discharge current. Under suitable conditions of pressure and current, the electrons in the discharge may directly excite the active atoms to produce a higher population in certain levels compared to lower levels. The relevant factors are the electron excitation cross section and the lifetimes of the various levels.

In the third method, an electric discharge is also employed. Here a suitable combination of gases is employed such that two different types of atoms, say A and B, both have some excited states, A^* and B^*, that coincide, or nearly coincide. In this case transfer of excitation may occur between the two atoms as follows:

$$A^* + B \rightarrow A + B^*$$

If the excited state of one of the atoms, say A^*, is metastable, then the presence of gas B will serve as an outlet for the excitation. As a consequence it is possible that the excited level of atom B may become more highly populated than some lower level to which the atom B can decay by radiation [Figure 9.3 (c)]. This is the case with the helium–neon laser. A neon atom receives its excitation from an excited helium atom. The laser transition then occurs in the neon atom.

The fourth method defines a class of lasers known as *chemical lasers*. Here a molecule is caused to undergo a chemical change in which one of the products of the reaction is a molecule, or an atom, that is left in an excited state. Under appropriate conditions a population inversion can occur. An example is the hydrogen fluoride chemical laser in which excited hydrogen fluoride molecules result from the reaction

$$H_2 + F_2 \rightarrow 2HF$$

9.5 Laser Oscillation

The optical cavity or resonator of a laser usually consists of two mirrors, curved or plane, between which the amplifying medium is located (Figure 9.4). If a sufficient population inversion exists in the

Figure 9.4. Basic laser setup.

medium, then the electromagnetic radiation builds up and becomes established as a standing wave between the mirrors. The energy is usually coupled from the resonator by having one or both of the mirrors partially transmitting.

In the case of plane reflectors, the optical cavity is similar to a conventional Fabry-Perot interferometer. The pass bands of the Fabry-Perot resonator occur at an infinite number of equally spaced frequencies

$$\dots, \nu_n, \nu_{n+1}, \nu_{n+2}, \dots$$

They differ by the free spectral range

$$\nu_{n+1} - \nu_n = \frac{c}{2d}$$

where c is the speed of light and d is the spacing of the reflectors. These frequencies define what are known as the *longitudinal modes* of the resonator. There are also transverse modes, as discussed in the following section.

Oscillation may occur at one or more of these resonant frequencies, depending on the width of the gain curve in relation to the mode spacing (Figure 9.5). Most lasers oscillate on several modes at once.

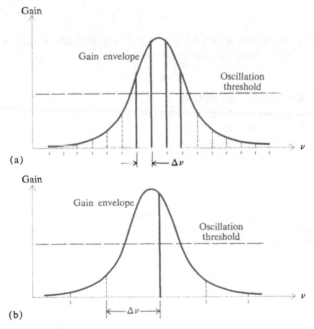

(a)

(b)

Figure 9.5. Oscillation frequencies in a laser. (a) Four longitudinal modes; (b) one mode.

If extremely high spectral purity is needed, however, it is possible to obtain oscillation on one mode by suitable selection of laser parameters. The inherent linewidth in this case is determined mainly by the quality factor Q of the laser resonator. These linewidths are typically of the order of a few hertz. However, in practice, linewidths of the order of 10^3 Hz are obtained. The limitation is determined largely by mechanical and thermal stability.

Threshold Condition for Oscillation We have seen that the irradiance of a parallel beam in an amplifying medium grows according to the equation

$$I_\nu = I_{0\nu} e^{\alpha_\nu\, x}$$

Suppose a wave in a laser cavity starts out at some point and travels back and forth between the cavity mirrors. On its return the wave will lose a certain fraction δ of its energy by scattering, reflection loss, and so forth. In order for the laser to oscillate, the gain must equal or exceed the loss, that is,

$$I_\nu - I_{0\nu} \geq \delta I_\nu$$

or, equivalently,

$$e^{\alpha_\nu 2l} - 1 \geq \delta \tag{9.17}$$

Here l is the active length of the amplifying medium. If $\alpha_\nu 2l \ll 1$, then the condition for oscillation can be written

$$\alpha_\nu 2l \geq \delta \tag{9.18}$$

If at a given frequency the gain exceeds the loss, the ensuing oscillation grows until an equilibrium condition is attained. The fractional loss δ is essentially constant and independent of the amplitude of the oscillation. Hence a depletion of the medium occurs that diminishes the population difference $N_2 - N_1$. The gain then drops until

$$\alpha_\nu 2l = \delta \tag{9.19}$$

The depletion occurs in a band centered on the oscillation frequency and is called *hole burning*. The shape of the hole is an inverted resonance curve, similar to the resonance curve of a harmonic

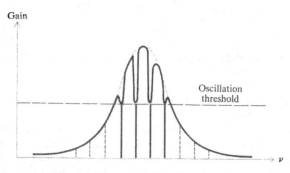

Figure 9.6. Hole burning of the gain envelope in a laser.

oscillator known as a *Lorentz profile*. The width of the Lorentz profile is equal to the reciprocal of the radiative lifetime of the lasing atom. If this radiative width is as large or larger than the width of the gain curve, then all of the excited atoms can be said to be "in communication with" the oscillating laser mode in question. This situation is termed *homogeneous broadening*. On the other hand if the radiative width of the laser transition is smaller than the width of the gain curve, then only part of the atoms participate in a given mode. This is called *inhomogeneous broadening*. In this case hole burning results in a modification of the gain curve as shown in Figure 9.6.

9.6 Optical-Resonator Theory

The concept of spatial modes of electromagnetic radiation in a closed cavity was briefly discussed in Section 7.3. There it was shown that a given mode can be specified by three integers that are directly related to the standing-wave pattern of the mode in question. In the case of a laser resonator, the cavity is not closed, being formed by only two reflecting surfaces. However, such open-sided resonators can still support a three-dimensional standing wave, sometimes referred to as a "quasi mode." One important fact is that part of the energy "spills" around the reflecting mirrors and is lost. This is called the *diffraction loss* of the resonator. A careful examination of such losses is necessary in laser applications, especially for low-gain systems such as the helium–neon laser in which the amplification per pass is typically only a few percent.

To illustrate the mathematical problem involved in the study of the optical resonator, we show in Figure 9.7 the resonator mirrors with aperture coordinates x,y and x',y', respectively. The case is equivalent to diffraction with multiple apertures, as shown. If $U(x,y)$ and $U'(x',y')$ represent the complex amplitudes of the radiation over the mirror surfaces, then by applying the Fresnel-Kirchhoff diffraction theory (Section 5.2) we can write

$$U'(x',y') = \frac{-ik}{4\pi} \iint U(x,y) \frac{e^{ikr}}{r} (1 + \cos \theta) \, dx \, dy \qquad (9.20)$$

in which

$$r = [d^2 + (x' - x)^2 + (y' - y)^2]^{1/2}$$

$$\cos \theta = \frac{d}{r}$$

If the mirrors are identical, which is typical, then for the steady state condition, that is, after the radiation has been reflected back and forth many times, the two functions U and U' will become identical,

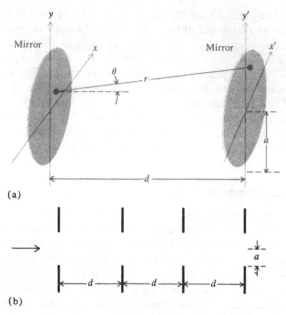

(a)

(b)

Figure 9.7. (a) Geometry of a Fabry-Perot laser cavity; (b) equivalent multiple diffraction problem.

except for a constant factor γ. In this case

$$\gamma U(x',y') = \iint U(x,y) \, K(x,y,x',y') \, dx \, dy \tag{9.21}$$

where

$$K(x,y,x',y') = \frac{-ik}{4\pi} (1 + \cos \theta) \frac{e^{ikr}}{r} \tag{9.22}$$

Equation (9.21) is an integral equation in the unknown function U. The function K is called the *kernel* of the equation, and γ is known as the *eigenvalue*. There are an infinite number of solutions U_n, $n = 1,2, \ldots$, each with an associated eigenvalue γ_n. The various solutions correspond to the normal modes of the resonator. Expressing γ_n as

$$\gamma_n = |\gamma_n| \, e^{i\phi_n} \tag{9.23}$$

we see that $|\gamma_n|$ specifies the ratio of the amplitude, and ϕ_n gives the phase shift associated with a given mode. The quantity $1 - |\gamma_n|^2$ is the relative energy loss per transit due to diffraction. (This is in addition to losses caused by absorption by the mirrors.)

Fox and Li were among the first to study the integral equation of the Fabry-Perot resonator [9]. They employed an electronic digital computer to find numerical solutions and associated eigenvalues. Boyd and Gordon found analytical solutions [6]. The types of optical resonators studied included both plane-parallel mirrors and curved mirrors.

Although accurate solutions of the Fabry-Perot resonator problem are quite involved, it is possible to find a simple approximation by employing the same procedure as that of the Fraunhofer diffraction case. Equation (9.22) then reduces to

$$K(x,y,x',y') = Ce^{-ik_1(xx'+yy')}$$

in which, under the approximation cited, C and k_1 are constants. The integral equation (9.21) then becomes

$$\gamma U(x',y') = C \iint U(x,y)e^{ik_1(xx'+yy')} \, dx \, dy \qquad (9.24)$$

This equation says that the function $U(x,y)$ is its own Fourier transform. The simplest of such functions is the Gaussian

$$U(x,y) = e^{-\rho^2/w^2} = e^{-(x^2+y^2)/w^2} \qquad (9.25)$$

Here w is a scaling constant, and $\rho^2 = x^2 + y^2$. More general functions that are their own Fourier transforms are products of functions known as *Hermite polynomials.* [27] and the Gaussian above, namely,

$$U_{pq}(x,y) = H_p\left(\frac{\sqrt{2}x}{w}\right) H_q\left(\frac{\sqrt{2}y}{w}\right) e^{-(x^2+y^2)/w^2} \qquad (9.26)$$

The integers p and q denote the order of the Hermite polynomials,[3] and each set (p,q) corresponds to a particular transverse mode of the resonator.

The lowest-order Hermite polynomial H_0 is a constant, namely, unity, hence the simple Gaussian mode corresponds to the set $(0,0)$ and is called the $TEM_{0,0}$ mode. The terminology TEM refers to the transverse electromagnetic waves in the cavity. Sometimes the notation $TEM_{n,p,q}$ is used, where the integer n is the longitudinal-mode number, and p and q are the transverse-mode numbers. Some low-order mode patterns are illustrated in Figure 9.8.

[3] The Hermite polynomials are

$$H_0(u) = 1$$
$$H_1(u) = 2u$$

$$H_n(u) = (-1)^n e^{u^2} \frac{d^n}{du^n} e^{-u^2}$$

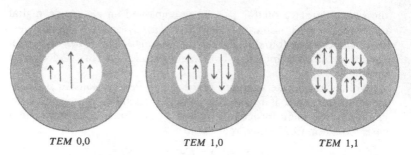

TEM 0,0   TEM 1,0   TEM 1,1

Figure 9.8. Field distributions at the mirrors for some low-order modes.

Resonator Configurations. Stability There are many combinations of curved and plane mirrors that can be used for laser cavities. A few of these are shown in Figure 9.9. One of the most commonly used cavity

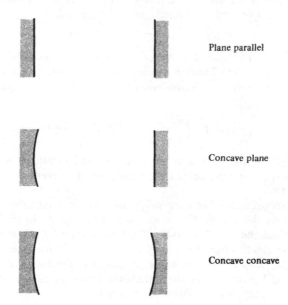

Plane parallel

Concave plane

Concave concave

Figure 9.9. Some common laser cavities.

configurations is known as the *confocal resonator*. This consists of two identical concave spherical mirrors separated by a distance equal to the radius of curvature. The confocal cavity is very much easier to align than the plane-parallel type. The latter requires an adjustment accuracy of the order of one arc second, whereas the confocal con-

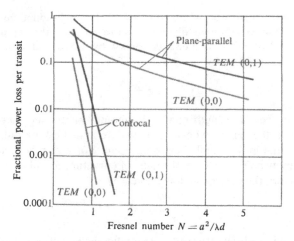

Figure 9.10.   Loss curves for the first two modes in plane-parallel and confocal laser cavities.

figuration, being somewhat self-aligning, needs a setting accuracy of only about a quarter of a degree in a typical application.

The diffraction loss, as calculated by Boyd and Gordon, for some of the low-order modes in plane-parallel and confocal resonators is plotted in Figure 9.10. The loss is plotted as a function of the *Fresnel number* $N = a^2/\lambda d$, where a is the mirror radius and d is the mirror separation. With confocal spherical mirrors, the diffraction losses of low-order modes are negligibly small when $N > 1$. A comparison of the losses for the plane-parallel resonator and the confocal resonator shows that the latter is definitely superior.

By the use of geometrical optics it is possible to classify laser resonators according to a criterion known as *stability*. A stable resonator is one in which a ray inside the cavity will remain close to the optic axis upon multiple reflections between the end mirrors. Stability criteria will be derived in the next chapter (Section 10.5). As we shall show, for the symmetrical cavity consisting of two mirrors of the same radius of curvature, the distance of separation must be less than twice the radius of curvature to form a stable cavity.

Spot Size   The scale parameter w, introduced in Equations (9.25) and (9.26), is a measure of the lateral distribution of the energy in the optical beam inside the resonator. The Gaussian function $e^{-\rho^2/w^2}$ falls to e^{-1} when ρ, the lateral distance from the optic axis, is equal to w. This function is proportional to the field amplitude; so the energy, being proportional to the square of the field, will fall to e^{-2} of its maximum value. Hence w is called the "spot size" of the dominant $(0,0)$ mode.

In a given resonator, w is a function of the longitudinal position. If we call z the longitudinal distance measured from the midpoint between the two mirrors, then, as shown by Boyd and Gordon, the parameter w is given by

$$w^2 = w_0^2 + \frac{\lambda^2 z^2}{\pi^2 w_0^2} \tag{9.27}$$

Here λ is the wavelength and w_0 is another parameter, namely, the spot size at the center, whose value is determined by the radii of curvature of the mirrors and their separation. For a symmetrical cavity formed by two mirrors each of radius of curvature R and separated by a distance d, the parameter w_0 is given by

$$w_0^2 = \frac{\lambda}{\pi} \left[\frac{d}{2} \left(R - \frac{d}{2} \right) \right]^{1/2} \tag{9.28}$$

and the radius of curvature of the standing-wave surfaces is given by

$$r_c = z + \frac{d(2R - d)}{4z} \tag{9.29}$$

In the case of a confocal resonator $R = d$, we find

$$w_0 = \sqrt{\frac{\lambda d}{2\pi}}$$

for the spot size at the center, and

$$w = \sqrt{\frac{\lambda d}{\pi}}$$

for the spot size at either mirror, $z = \pm d/2$. Figure 9.11 illustrates the

Figure 9.11.   Standing wave pattern and lateral distribution of the $TEM_{0,0}$ mode of a confocal laser cavity.

confocal case. Surfaces of constant phase are drawn to show the curvature of the standing waves within the cavity. At the mirrors the wave surfaces match the curvature of the mirror surface. At the center, where the spot size is minimum, the wave surface becomes planar. A hemiconfocal cavity would be represented by placing a plane mirror at the center of the confocal cavity, the wave surfaces and spot size remaining unchanged for the portion of the confocal cavity that lies between the mirrors forming the new cavity. In fact, *any* two wave surfaces will define a cavity if the wave surfaces are replaced by mirrors that match the curvatures of the wave surfaces.

9.7 Gas Lasers

Figure 9.12 shows a typical physical arrangement of a gas laser. The optical cavity is provided by external mirrors. The mirrors are coated with multilayer dielectric films in order to obtain high reflectance at the desired wavelength. Spherical mirrors arranged in the confocal configuration are used because of the low loss and ease of adjustment of this type of cavity.

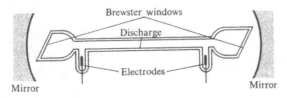

Brewster windows

Discharge

Electrodes

Mirror   Mirror

Figure 9.12. Typical design of a gas laser.

The laser tube is fitted with Brewster end windows in order to obtain the maximum possible transparency. When Brewster windows are used, the output of the laser is linearly polarized. The reason is that the windows are highly transparent for one preferred direction of polarization—the *TM* polarization—as discussed in Section 2.8. Consequently, laser oscillation at this favored polarization builds up and becomes dominant over the orthogonal *TE* polarization.

External electrical excitation may be provided in any of the following ways:

(1) Direct-current discharge
(2) Alternating-current discharge
(3) Electrodeless high-frequency discharge
(4) High-voltage pulses

Methods (1) and (2) are commonly employed in commercial gas lasers. The direct-current discharge (1) is advantageous if the laser is to be used for such things as optical heterodyning, communications, and so forth. The alternating-current discharge (2) is simplest because the power source need be only an ordinary high-voltage transformer connected to cold metal electrodes in the tube. The electrodeless high-frequency discharge (3) was used in the first gas laser—the helium–neon laser—developed at the Bell Telephone Laboratories by

Javan, Bennett, and Herriott [21]. Method (4) is employed in many high-power pulsed lasers and is necessary in some instances in which a steady population inversion cannot be maintained.

The Helium–Neon Laser Figure 9.13 shows an energy-level diagram of the helium–neon laser system. Helium atoms are excited by elec-

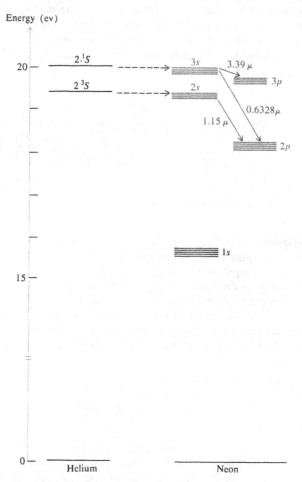

Figure 9.13. Simplified energy-level diagram of the helium–neon laser.

tron impact in the discharge. The populations of the metastable states ^3S and ^1S of helium build up because there are no optically allowed transitions to lower levels. It is seen from the figure that the neon levels labeled $2s$ and $3s$ lie close to the metastable helium levels.

There is, therefore, a high probability of energy transfer when a metastable helium atom collides with an unexcited neon atom. These energy transfers are

$$He(^3S) + Ne \rightarrow He + Ne(2s)$$
$$He(^1S) + Ne \rightarrow He + Ne(3s)$$

Under suitable discharge conditions a population inversion of the $Ne(2s)$ and $Ne(3s)$ levels can take place. The optimum value of total pressure is about 1 torr, and the most favorable ratio of helium to neon is found to be about $7:1$.

The main laser action in the helium–neon system corresponds to the following transitions in the neon atom:

$$3s_2 \rightarrow 2p_4 \ 632.8 \text{ mm}$$
$$2s_2 \rightarrow 2p_4 \ 1.1523 \ \mu$$
$$3s_2 \rightarrow 3p_4 \ 3.39 \ \mu$$

In addition to these, many weaker transitions in neon have been made to undergo laser oscillation [25].

Other Gas Lasers Electric discharges in pure gases and in various mixtures have produced laser action at a large number of different wavelengths from the far infrared to the ultraviolet. All of the noble gases, helium, neon, argon, krypton, and xenon exhibit laser transitions in the pure gas. The argon ion laser, for example, generates visible light at several wavelengths in the blue region. Pulsed discharges in metal vapors have been used to obtain laser action in zinc, cadmium, mercury, lead, tin, and other metals [4] [8] [37]. The halogens, chlorine, bromine, and iodine, similarly yield laser transitions under pulsed conditions [22]. Discharges in molecular gases have yielded population inversions on transitions in various molecules. Notable among these are the molecular nitrogen laser (N_2) that produces infrared and ultraviolet radiation, and the CO_2 laser that oscillates in the 10 μ region. Table 9.1 summarizes some of the common gas lasers.

9.8 Optically Pumped Solid-State Lasers

In solid-state lasers, of which ruby is the prototype, the active atoms of the laser medium are embedded in a solid. Both crystals and glasses have been used as the supporting matrices for this application. The crystal or glass is usually made in the form of a cylindrical rod whose ends are optically ground and polished to a high degree of parallelism and flatness. The laser rod may be made to constitute its own optical cavity by coating the ends, or external mirrors may be employed.

Table 9.1. SOME GAS LASERS

Gas or Gas Mixture	Active Species	Principal Laser Wavelengths (μ)	Remarks
He-Ne	Ne	0.6328, 1.15, 3.39	cw*
Ne	Ne	0.5401, 0.6143, 1.15	pulsed high gain
Ne	Ne$^+$	0.3323, 0.3378, 0.3392	cw or pulsed
Ar	Ar$^+$	0.4765, 0.4880, 0.5145	cw or pulsed
Kr	Kr$^+$	0.5208, 0.5309, 0.5862, 0.6471	cw or pulsed
Xe-He	Xe	3.507, 5.574	cw high gain
Xe	Xe$^+$	0.4603, 0.5419, 0.5971	cw or pulsed
Ne-O$_2$, Ar-O$_2$	O	0.8446	cw
N$_2$	N$_2$	0.3371	pulsed high gain
Air, N$_2$	N$^+$	0.5679	pulsed
He-Cd	Cd$^+$	0.3250, 0.4416	cw high efficiency
CO$_2$-N$_2$-He	CO$_2$	10.6	pulsed and cw, high power and high efficiency
H$_2$	H$_2$	0.12, 0.17	first vacuum ultraviolet laser
H$_2$O	H$_2$O	27.9, 118.6	long wavelength
CH$_3$CN-NH$_3$	HCN	337	very long wavelength

* Continuous wave, cw.

Optical pumping of the active atoms is accomplished by means of an external light source. This source may be either pulsed or continuous. High-intensity lamps, such as xenon flash lamps or high-pressure mercury discharge lamps, are generally used for this purpose. Figure 9.14 shows two typical arrangements of optically pumped

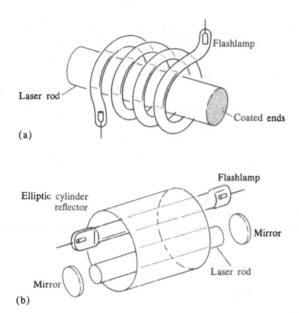

Figure 9.14. Typical designs for optically pumped solid-state lasers.

solid-state lasers. In (a) a helical flash lamp is used with the laser rod placed inside the helix. The rod ends are coated so that the complete laser is very simple and compact. A more elaborate system is shown in (b). The laser rod is placed at one focus of an elliptical reflector, and the pumping lamp is placed at the other focus. External mirrors are used for the optical cavity.

The Ruby Laser   The rod of a ruby laser is made of synthetic sapphire (Al_2O_3), which is doped with approximately 0.05 percent by weight of Cr_2O_3. This concentration produces a pink-colored material. The color is due to the presence of Cr^{3+} ions, which replace aluminum in the crystal lattice.

An energy-level diagram of Cr^{3+} in ruby is shown in Figure 8.17 of the previous chapter. In operation of the laser the pumping light is absorbed by the Cr^{3+} ions raising them from the ground state ^4A to either of the excited states $^4T_1$ or $^4T_2$. From these levels a rapid radiationless transition to the level ^2E takes place. Decay from ^2E is relatively slow, so that with sufficient excitation a population inversion between ^2E and the ground state ^4A can occur. When this condition attains, amplification occurs at the wavelength 6934 Å, corresponding to the transition $^2E \rightarrow {}^4A$. The resultant output is an intense pulse of light at this wavelength.

Other Solid-State Laser Materials Besides ruby, a number of other crystals, when doped with impurity atoms containing incomplete subshells, have been found to exhibit stimulated emission. Neodymium, for example, has a laser transition at about 1.06 μ. Various host solids have been used with neodymium, including crystals of calcium fluoride (CaF_2) and calcium tungstate ($CaWO_4$), as well as glass. Neodymium-doped yttrium aluminum garnet (YAG) has proved to be very efficient as a lasing crystal. YAG lasers can be operated continuously when pumped with incandescent tungsten lamps. Table 9.2 lists a number of examples of solid-state laser materials.

Table 9.2. SOME SOLID-STATE LASERS

Laser Ion	Host Material	Wavelength (μ)	Remarks
Cr^{3+}	Al_2O_3	0.6943	High power, first laser discovered
Nd^{3+}	CaF_2	1.046	
	$CaWO_4$	1.06	
	Glass	1.06	Widely used
	YAG	1.06	Efficient, capable of cw* operation
Ho^{3+}	CaF_2	2.09	
Er^{3+}	CaF_2, glass	2.04	
Tm^{3+}	SrF_2, $CaWO_4$	1.9	
Yb^{3+}	Glass	1.1	
U^{3+}	BaF_2, CaF_2	2.5	

* Continuous wave, cw.

9.9 Dye Lasers

Stimulated emission in liquid solutions of fluorescent organic dyes was first reported in 1966 by Sorokin and Lankard at the IBM laboratories. The dye solutions were optically pumped with a ruby laser, and in later experiments with a fast flash lamp. In the organic dyes used in lasers, such as fluorescein and rhodamine, the fluorescence bands are quite broad, typically 50 to 100 nm. The amplification bands are correspondingly broad, and it is thus possible to vary the output frequency of the laser by the use of tuning elements (prisms, gratings, and interferometers) inserted in the laser cavity.

Previous to the advent of the dye laser, only a finite number of discrete laser frequencies was available. Now, with a wide selection of lasing dyes, the entire region of the optical spectrum from the near ultraviolet to the near infrared can be spanned. Continuous-wave operation of the dye laser has been achieved by using a continuous-wave gas laser (argon or krypton) as the optical pump source. A diagram of a typical continuous-wave tunable dye laser is shown in Figure 9.15. Table 9.3 lists the tuning ranges of some laser dyes.

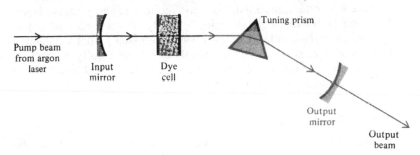

Figure 9.15.   A tunable dye laser arrangement. The input mirror is transparent to the argon laser pump beam, but is highly reflecting over the tuning range of the dye.

Table 9.3. SOME LASER DYES

Dye	Approximate Tuning Range (nm)
Cresyl violet	640–700
Rhodamine B	580–690
Acridine red	600–630
Rhodamine 6G	560–650
Sodium fluorescein	520–570
4 Methylumberiferone	440–540
Coumarin	440–490
Diphenyl anthracene	430–450
Sodium salicylate	390–420
POPOP	380–440

9.10 Semiconductor Diode Lasers

The most compact laser is the semiconductor diode laser, also called the *injection laser*. In its simplest form the diode laser consists of a *p-n* junction in a doped single crystal of a suitable semiconductor, such as gallium arsenide. When a forward bias is applied to the diode, electrons are injected into the *p* side of the junction and holes are injected into the *n* side. The recombination of holes and electrons within the junction region results in *recombination radiation*. This is the principle for the operation of light-emitting-diode (LED) devices that are used for many applications—such as numerical display in electronic computers—in which *spontaneous* recombination radiation is produced. If the junction current density is large enough, a population inversion can be obtained between the electron levels and the hole levels. Stimulated emission can then occur and laser action commences when the optical gain exceeds the loss in the junction layer. In diode lasers this layer is thin, typically of the order of a few microns, and the end faces of the crystal are made partially reflective to form an optical resonator. The threshold current density for gallium arsenide injection lasers is about 10^4 A/cm^2, and the emitted radiation is in the near infrared, about 830 to 850 nm.

9.11 Q-Switching and Mode Locking

In high-power lasers that are pumped by intense optical pulses the lasing action begins when the population-inversion density reaches a certain threshold value, namely, when the optical gain exceeds the loss. The reciprocal of the fractional loss per cycle is known as the *resonant Q* of the cavity. The lower the loss, the higher the Q and the lower the inversion density required for oscillation, and conversely. In high-power pulsed lasers the inversion density is quickly "used up" as soon as laser oscillation commences. By delaying the onset of oscillation it is possible to obtain a higher inversion and thus higher output than would otherwise be obtained. This is achieved by inserting a suitable optical shutter, known as a Q *switch,* into the laser cavity. The shutter is closed at the beginning of the pump pulse and opened when the population inversion is maximum.

There are various methods of Q switching. A simple method is to rotate one of the resonator mirrors at a high rate about an axis perpendicular to the optical axis of the laser cavity. This "spoils" the cavity and prevents oscillation except for a very brief time during the rotation cycle. A second method makes use of an electro-optic shutter in the laser cavity. Pockels cells are commonly used for this application. The Pockels cell is "opened up" with a high-voltage pulse that is appropriately delayed relative to the optical pump pulse. A third method employs what is known as a *saturable absorber*. This is a certain type of dye which bleaches out or becomes transparent when strongly irradiated so that the upper levels are saturated and thus no more absorption can occur. This is known as *passive Q* switching.

By the use of Q switching it is possible to attain very high peak powers. Pulses in the 100-MW range are common with ruby and neodymium-glass lasers. Also, the pulse duration is shortened from typical values of a few microseconds for conventional operation to a few nanoseconds for Q-switched operation. Even higher powers can be attained by passing the output beam of a Q-switched laser through an amplifier or a series of amplifiers. In this way it has been possible to produce pulses in the gigawatt (GW) and terawatt (TW) region. These extremely high powers are necessary for laser fusion.

Mode Locking If a nonlinear absorber, such as a bleachable dye, is placed in the resonator cavity of a laser, it is often found that the laser output is changed into very short, regularly spaced pulses. The time separation between successive pulses is found to be the round-trip time for light in the optical cavity (or sometimes a simple fraction, $\frac{1}{2}$, $\frac{1}{3}$, . . . of the round trip time). Thus the radiation inside the cavity has become "bunched" into one or more narrow pulses that bounce back and forth between the resonator mirrors. Now a simple Fourier analysis shows that the frequency spectrum of a regular pulse train consists of a number of discrete frequencies separated by the repetition frequency of the pulses. Since the round-trip time is $2d/u$, where d is the mirror spacing and u is the velocity, then the frequency spectrum of the pulse train consists of components separated by the longitudinal mode spacing $u/2d$ (or an integral multiple of it). However, unlike a conventional laser that is oscillating on several different longitudinal modes simultaneously but in a mutually incoherent phase relationship, the phases in the case of regular pulsing are necessarily related in a definite way. Hence the pulsing laser is said to be *mode locked* or *phase locked*.

Now the width of a single pulse is equal to the reciprocal of the total frequency bandwidth involved, according to the discussion of Section 4.6, Equation (4.33). For a mode-locked laser this is the total bandwidth occupied by the oscillating modes. Hence if there are N such modes, the temporal width of each pulse is the fraction $1/N$ of the time interval between successive pulses. Thus lasers having a broad frequency band of amplification can support a large number of modes and are therefore capable of producing very short mode-locked pulses.

In the case of gas lasers, such as the helium–neon or the argon laser, which oscillate on a narrow atomic line transition, the mode-locked pulses are limited in narrowness to about a few nanoseconds. On the other hand, with dye lasers and the neodymium-glass laser that have broadband capability, mode-locked pulses in the picosecond range are produced.

It is interesting to note that a 3-ps pulse measures only about 1 mm thick, as a typical example. The radiation in such cases can truly be said to be concentrated into a thin "sheet" that travels with the speed of light.

9.12 The Ring Laser

The ring laser is an example of a technological application of lasers. It is a device designed to measure rotation by means of counterrotating coherent light beams and was developed in 1963 at the Sperry-Rand Corporation. The device is the laser analogue of the Sagnac experiment (Section I.4 in the Appendix).

The basic design is shown in Figure 9.16. The optical cavity of the laser consists of four mirrors arranged in a square. One or more laser

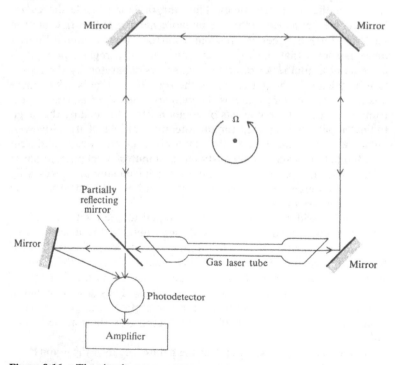

Figure 9.16. The ring laser.

tubes are inserted into the cavity to provide amplification. Oscillation occurs at those resonance frequencies

$$\nu_n = n\,\frac{c}{L}$$

which lie within the amplification curve of the laser medium. Here L is the effective length of the complete loop and n is an integer.

If the system is rotating about an axis perpendicular to the plane of the loop, the effective path lengths for the clockwise and the counterclockwise beams are different. This results in a difference $\Delta\nu$ between the frequencies of the laser oscillation of the two beams. The difference is

$$\Delta\nu = \nu\,\frac{4A}{c}\,\Omega$$

where ν is the frequency of the laser when it is not rotating, A is the loop area, and Ω is the angular speed of rotation.

In the operation of the ring laser, the method of optical heterodyning is used. The two output beams of the laser are brought together and fed to a photodetector as shown. The output of the photodetector consists of a beat signal whose frequency is equal to the difference $\Delta\nu$. This, in turn, is proportional to the angular speed of rotation.

PROBLEMS

9.1 The first line of the principal series of sodium is the D line at 590 nm. This corresponds to a transition from the first excited state ($3p$) to the ground state ($3s$). What is the energy in electron volts of the first excited state?

9.2 What fraction of sodium atoms are in the first excited state in a sodium vapor lamp at a temperature of 250°C?

9.3 What is the ratio of stimulated emission to spontaneous emission at a temperature of 250°C for the sodium D line?

9.4 Calculate the gain constant of a hypothetical laser having the following parameters: inversions density $= 10^{17}/cm^3$, wavelength $= 700$ nm, line width $= 1$ nm, spontaneous emission lifetime $= 10^{-4}$ s.

9.5 Calculate the inversion density $N_2 - N_1\,(g_2/g_1)$ for a helium-neon laser operating at 633 nm. The gain constant is 2 percent/m, and the temperature of the discharge is 100°C. The lifetime of the upper state against spontaneous emission to the lower state is 10^{-7} s.

9.6 If the spot size at the mirrors of a helium-neon laser is 0.5 mm,

what is the length of the laser cavity? The cavity is of the con-
focal-type, and the wavelength is 633 nm. What is the spot size
of the 3.39-μm transition in the same cavity?

9.7   Limiting apertures are placed at the mirrors of a confocal cavity
in order to suppress the higher modes. If the cavity is 1 m in
length, what should the diameter of the apertures be in order
that the loss for the *TEM* 0,1 mode be 1 percent? What is the
corresponding loss for the *TEM* 0,0 mode? (See Figure 9.10.)
The wavelength is 633 nm.

9.8   Prove Equation (9.30) giving the difference frequency of the ring
laser.

CHAPTER 10
Ray Optics

10.1 Reflection and Refraction at a Spherical Surface

Virtually all of the optical instruments in common use contain only two types of optical surface, namely, plane and spherical. When light passes through an optical system containing various optical elements such as mirrors, prisms, and lenses, the path of the light is conveniently analyzed in terms of light rays. The trajectory of a ray can be computed by using the law of reflection or Snell's law of refraction. The procedure of following the path of a chosen ray through the system is known as *ray tracing*.

In Figure 10.1 is shown the path of a ray originating at a point P on the axis of a single optical surface, namely, a spherical mirror (a) or a spherical refracting surface separating two optical media (b). The ray returns to the axis at a certain point Q. The distance $OP = s$

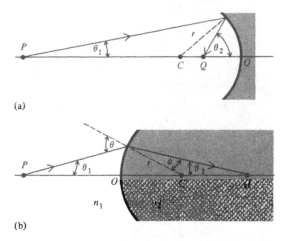

(a)

(b)

Figure 10.1. Geometry for (a) reflection of a ray by a spherical surface, and (b) refraction of a ray by a spherical surface.

294

is called the *object distance,* and the distance $OQ = s'$ is called the *image distance.* The radius of curvature of the surface is $OC = AC = r.$

In the case of the spherical mirror shown in Figure 10.1(a) from the law of sines in the appropriate triangles we have

$$r \sin \theta = (s - r) \sin \theta_1 \qquad r \sin \theta' = (r - s') \sin \theta_2$$

By use of the law of reflection $\theta = \theta'$, we find

$$\frac{\sin \theta_1}{\sin \theta_2} = \frac{(r - s')}{(s - r)} \tag{10.1}$$

for the relationship between the slope angles of the incident and the reflected rays.

Similarly, for the spherical refracting surface of Figure 10.1(b), the law of sines gives

$$r \sin \theta = (r + s) \sin \theta_1 \qquad r \sin \phi = (s' - r) \sin \theta_2$$

By using Snell's law, $\sin \theta / \sin \phi = n_2/n_1$, in which n_1 and n_2 are the indices of refraction of the two media, we obtain

$$\frac{\sin \theta_1}{\sin \theta_2} = \frac{s' - r}{s + r} \frac{n_2}{n_1} \tag{10.2}$$

for the relationship between the incident and the refracted rays.

Paraxial Approximation   When a bundle of rays originates from an axial point, the analysis given above shows that the image distances are not the same for all rays but rather are functions of the original slope angles θ_1 at the object point. This means that the rays do not come to a single focus. This is a common feature of spherical reflecting and refracting surfaces and is known as *spherical aberration.* However, if the angles are small enough for the sines to be replaced by the angles themselves in Equations (10.1) and (10.2), an important simplification results. This is called the *paraxial approximation.* Thus in the case of the spherical reflector we find that Equation (10.1) reduces to

$$\frac{1}{s} + \frac{1}{s'} = \frac{2}{r} \tag{10.3}$$

and, similarly, for the spherical refracting surface we find that Equation (10.2) reduces to

$$\frac{n_1}{s} + \frac{n_2}{s'} = \frac{n_2 - n_1}{r} \tag{10.4}$$

Sign Convention It should be noted that although the diagrams in Figure 10.1 depict a ray that diverges from an axial object point and converges to an axial image point, the derived results. Equations (10.3) and (10.4), are valid for other cases. Namely, if it turns out that the image distance q is negative for given values of the other parameters, then the rays do not actually converge, but diverge and appear to come from a fictitious point behind the optical surface in question. Furthermore, in the case of a curved reflector, the sign of the radius r is positive for a concave surface, as shown, and negative if the surface is convex. Similarly, a curved refracting surface is considered to have a positive radius of curvature if it is concave toward the direction of ray travel, as shown, and negative if it is convex toward the direction of ray travel.

Focal Length of a Spherical Mirror Referring to Equation (10.3) we see that if the object distance is infinitely large, so that the incoming rays are parallel, the resulting image distance is just $r/2$. This distance is called the *focal length* of the mirror, and is denoted by f. We can then write

$$\frac{1}{s} + \frac{1}{s'} = \frac{1}{f} \tag{10.5}$$

in which

$$f = \frac{r}{2} \tag{10.6}$$

for a spherical mirror of radius r.

10.2   Lenses

Single Thin Lens For a single thin lens consisting of a material of index n with radii of curvature r_1 and r_2, it is easy to show that the object and image distances for paraxial rays are related by the equation

$$\frac{1}{s} + \frac{1}{s'} = (n - 1)\left(\frac{1}{r_1} - \frac{1}{r_2}\right) \tag{10.7}$$

The index of the surrounding material is considered here to be unity. If the lens is in a homogeneous medium, n is then the relative index of refraction of the lens material and the medium. The derivation of the above formula is left as an exercise. As in the case of the spherical reflector, we define the focal length f as the image distance for parallel incoming rays. Hence we can write

$$\frac{1}{f} = (n - 1)\left(\frac{1}{r_1} - \frac{1}{r_2}\right) \tag{10.8}$$

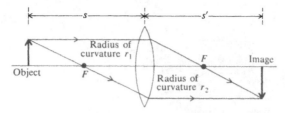

Figure 10.2. Formation of an image by a thin lens. The distance from the center of the lens to either of the points marked F is equal to the focal length f.

This equation is known as the *lens-maker's formula*. The formation of an image by a single lens is shown in Figure 10.2. Although the object and image points are not axial in this case, the paraxial formulas obtained above are still useful for obtaining approximate results.

Combinations of Thin Lenses The combined focal length f of any number of thin lenses placed close together (in contact) is expressible as

$$\frac{1}{f} = \frac{1}{f_1} + \frac{1}{f_2} + \frac{1}{f_3} + \cdots \tag{10.9}$$

where $f_1, f_2, \ldots$ are the focal lengths of the individual lenses. For two lenses not in contact but separated by a distance d, the effective focal length of the combination is given by the expression

$$\frac{1}{f} = \frac{1}{f_1} + \frac{1}{f_2} - \frac{d}{f_1 f_2} \tag{10.10}$$

Thick Lens For a single thick lens in air, Equation (10.2) applies where, however, the object and image distance are measured from the *principal planes*, H and H' (Figure 10.3). The focal length f is calculated from the equation

$$\frac{1}{f} = (n - 1) \left[\frac{1}{r_1} + \frac{1}{r_2} - \frac{(n - 1)^2 t}{n r_1 r_2} \right] \tag{10.11}$$

where t is the lens thickness. The positions of the principal planes are given by

$$d_1 = ft \left(\frac{1 - n}{r_2} \right)$$

$$d_2 = ft \left(\frac{1 - n}{r_1} \right) \tag{10.12}$$

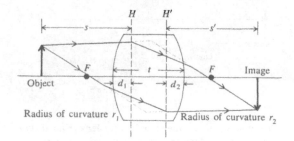

Figure 10.3.   Formation of an image by a thick lens. The focal length f, given by Equation 10.11, is the distance from either principal plane to the points marked F.

Chromatic Aberration   Due to dispersion, the focal length of a simple lens, Equation (10.8), varies with the wavelength. This variation, called *chromatic aberration*, can be reduced substantially by means of a lens combination in which the component lenses are made of glasses having different dispersions. An *achromatic* combination of focal length f for two thin lenses in contact is obtained if the focal lengths of the component lenses are

$$f_1 = f\left(1 - \frac{\delta_1}{\delta_2}\right) \qquad f_2 = f\left(1 - \frac{\delta_2}{\delta_1}\right) \tag{10.13}$$

where

$$\delta_1 = \frac{1}{n_1 - 1}\frac{dn_1}{d\lambda} \qquad \delta_2 = \frac{1}{n_2 - 1}\frac{dn_2}{d\lambda} \tag{10.14}$$

Since $dn/d\lambda$ also varies with wavelength, a lens can be achromatized over a limited wavelength interval only.

Spherical Aberration of a Single Lens   In the case of a simple lens, the effective focal length varies with the distance h at which the incident rays enter the lens (Figure 10.4). For *paraxial* rays formula (10.8) applies. The difference Δ between the focal length f for paraxial rays

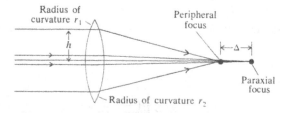

Figure 10.4.   Illustrating spherical aberration.

and the focal length for rays entering the lens a distance h from the axis is approximately proportional to h^2. It can be expressed

$$\Delta = \tfrac{1}{2} K h^2 \tag{10.15}$$

in which K is given by the rather complicated expression

$$K = \frac{f^2(n-1)}{n^2} \left[\frac{1}{r_1{}^3} + \left(\frac{1}{f} + \frac{1}{r_2} \right) \left(\frac{n+1}{f} + \frac{1}{r_2} \right) \right] \tag{10.16}$$

This is minimum when the ratio of the radii is

$$\frac{r_1}{r_2} = \frac{n+4-2n^2}{n+2n^2} \tag{10.17}$$

The above formula giving the shape of a simple lens for minimum spherical aberration holds only if the object is at an infinite distance.

In optical instruments that use combinations of lenses, such as cameras, spherical aberration and other aberrations can be minimized by appropriate choice of lens curvatures and separations. The mathematical complexities of lens design have been greatly facilitated in recent years by the use of electronic computers. The reader who is interested in this subject is referred to the textbook list at the end of the book.

10.3 Ray Equations

Consider a set of paraxial rays traveling in the general direction of the optic axis (z axis) of an optical system. The position and direction of any ray can be defined by two parameters, namely, the distance from the optic axis, which we shall call ρ, and the angle the ray makes with the optic axis, which we shall denote by θ. In particular, if the ray proceeds from $z = z_1$ to $z = z_2 = z_1 + d$, as shown in Figure 10.5, then in the paraxial approximation, the value of ρ increases by θd, and so we can write

$$\rho_2 = \rho_1 + \theta_1 d \tag{10.18}$$
$$\theta_2 = \theta_1$$

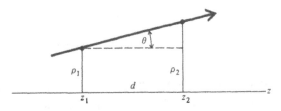

Figure 10.5. Geometry defining the parameters of the ray vector.

Equation (10.18) expresses the transformation of a ray (ρ_1, θ_1) when it travels a distance d in a single homogeneous medium.

Next, suppose a given ray (ρ_1, θ_1) passes through a plane dielectric interface separating two media of indices n_1 and n_2 (Figure 10.6).

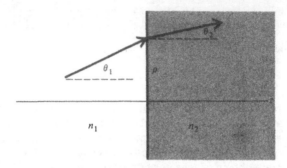

Figure 10.6.   Geometry of a ray passing through a plane interface.

Then, again in the paraxial approximation, we have from Snell's law

$$\rho_2 = \rho_1$$
$$\theta_2 = \theta_1 \, \frac{n_1}{n_2} \qquad\qquad (10.19)$$

If instead of a plane boundary we have a curved dielectric interface, as was shown in Figure 10.1(b), then with an obvious change in the notation we have

$$\rho_2 = \rho_1$$
$$\theta_2 = \theta_1 \frac{n_1}{n_2} - \left(1 - \frac{n_1}{n_2}\right) \frac{\rho_1}{r} \qquad (10.20)$$

which follows from Equation (10.4) by setting $\theta_1 = \rho_1/s$, $\theta_2 = -\rho_1/s'$.

Finally, in the case of a lens or a curved mirror, the ray equations corresponding to Equation (10.5) are

$$\rho_2 = \rho_1$$
$$\theta_2 = \theta_1 - \frac{\rho_1}{f} \qquad\qquad (10.21)$$

in which f is the focal length of the lens or mirror.

10.4   Ray Matrices and Ray Vectors

The ray equations of the preceding section can all be conveniently expressed in a general matrix form

$$\begin{bmatrix} \rho_2 \\ \theta_2 \end{bmatrix} = \begin{bmatrix} M_{11} & M_{12} \\ M_{21} & M_{22} \end{bmatrix} \begin{bmatrix} \rho_1 \\ \theta_1 \end{bmatrix} \qquad (10.22)$$

The column matrices $\begin{bmatrix} \rho_1 \\ \theta_1 \end{bmatrix}$ and $\begin{bmatrix} \rho_2 \\ \theta_2 \end{bmatrix}$ are called *ray vectors*, and the transformation matrix

$$\begin{bmatrix} M_{11} & M_{12} \\ M_{21} & M_{22} \end{bmatrix}$$

is known as the *ray matrix*. It characterizes in a compact way the change that a ray undergoes in passing through an optical surface or element. Table 10.1 lists the ray matrices that correspond to the ray equations derived in Section 10.3.

The most useful aspect of the matrix treatment of ray optics is that combinations of optical elements are easily handled by matrix multiplication. Thus the overall ray matrix for a series of optical elements whose individual matrices are $[M_a]$, $[M_b]$, $[M_c]$, and so on, is given by

$$[M] = \cdots [M_c][M_b][M_a] \qquad (10.23)$$

Table 10.1. SOME RAY MATRICES

Optical Element	Ray Matrix	Comments
Free travel through homo-geneous medium	$\begin{bmatrix} 1 & d \\ 0 & 1 \end{bmatrix}$	$d =$ distance of travel
Plane dielectric interface	$\begin{bmatrix} 1 & 0 \\ 0 & n_1/n_2 \end{bmatrix}$	n_1 and n_2, indices of refraction
Curved dielectric interface	$\begin{bmatrix} 1 & 0 \\ \frac{1}{r}\left(\frac{n_1}{n_2} - 1\right) & \frac{n_1}{n_2} \end{bmatrix}$	$r =$ radius of curvature convex $r > 0$ concave $r < 0$
Thin lens or mirror of focal length f	$\begin{bmatrix} 1 & 0 \\ -1/f & 1 \end{bmatrix}$	$f = r/2$ for mirror Use Equation (10.8) for lens

For example, consider a combination of two thin lenses held close together, of focal lengths f_1 and f_2, respectively. Referring to Table 10.1 for the ray matrix of a single lens, we see that the matrix for the combination is given by

$$\begin{bmatrix} 1 & 0 \\ -\dfrac{1}{f_2} & 1 \end{bmatrix} \begin{bmatrix} 1 & 0 \\ -\dfrac{1}{f_1} & 1 \end{bmatrix} = \begin{bmatrix} 1 & 0 \\ -\left(\dfrac{1}{f_1} + \dfrac{1}{f_2}\right) & 1 \end{bmatrix}$$

hence the effective focal length of the combination is expressed by

$$\frac{1}{f} = \frac{1}{f_1} + \frac{1}{f_2}$$

10.5 Periodic Lens Waveguides and Optical Resonators

The method of ray matrices will now be employed to analyze the behavior of a ray that passes through a series of many lenses. Also, since a curved mirror of radius r is optically equivalent to a lens of focal length $f = r/2$, the same analysis applies to a ray that is reflected back and forth between two curved mirrors. For simplicity we consider the symmetrical case in which the lenses (mirrors) are identical, of focal length f, and uniformly spaced a distance d apart as shown in Figure 10.7.

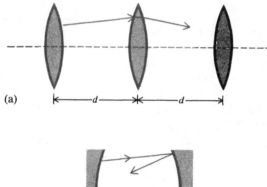

(a)

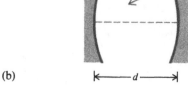

(b)

Figure 10.7. Symmetrical periodic lens waveguide (a), and optical resonator (b).

Consider a ray that starts out at the surface of one lens or mirror, travels a distance d to the next lens or mirror, and is refracted by the lens or reflected by the mirror. The transformation of the ray is then given by the following equation:

$$\begin{bmatrix} \rho_2 \\ \theta_2 \end{bmatrix} = \begin{bmatrix} 1 & 0 \\ -\dfrac{1}{f} & 1 \end{bmatrix}\begin{bmatrix} 1 & d \\ 0 & 1 \end{bmatrix}\begin{bmatrix} \rho_1 \\ \theta_1 \end{bmatrix} = \begin{bmatrix} 1 & d \\ -\dfrac{1}{f} & 1 - \dfrac{d}{f} \end{bmatrix}\begin{bmatrix} \rho_1 \\ \theta_1 \end{bmatrix} \qquad (10.24)$$

At this point we consider the following question: Is there any initial ray vector such that the output ray vector is equal to the initial ray vector multiplied by a constant factor? In other words, do there exist any solutions of the equation

$$\begin{bmatrix} \rho_2 \\ \theta_2 \end{bmatrix} = \lambda \begin{bmatrix} \rho_1 \\ \theta_1 \end{bmatrix}$$

The same question could be asked in the case of any optical system having an overall ray matrix as expressed by Equation (10.22). The mathematical problem is one of finding solutions to the equation

$$\begin{bmatrix} M_{11} & M_{12} \\ M_{21} & M_{22} \end{bmatrix}\begin{bmatrix} \rho_1 \\ \theta_1 \end{bmatrix} = \lambda \begin{bmatrix} \rho_1 \\ \theta_1 \end{bmatrix}$$

or, equivalently,

$$\begin{bmatrix} M_{11} - \lambda & M_{12} \\ M_{21} & M_{22} - \lambda \end{bmatrix}\begin{bmatrix} \rho_1 \\ \theta_1 \end{bmatrix} = 0 \qquad (10.25)$$

We have here the same sort of mathematical problem as that discussed earlier in Chapter 2 in connection with the eigenvectors of Jones matrices. Solutions of Equation (10.25) define the eigenvectors optical system. The secular determinant is

$$\begin{vmatrix} M_{11} - \lambda & M_{12} \\ M_{21} & M_{22} - \lambda \end{vmatrix} = 0 \qquad (10.26)$$

Returning to the problem of the symmetrical resonator or periodic waveguide, the secular determinant is

$$\begin{bmatrix} 1 - \lambda & d \\ -\dfrac{1}{f} & 1 - \dfrac{d}{f} - \lambda \end{bmatrix} = 0 \qquad (10.27)$$

which reduces to

$$\lambda^2 + \lambda \left(2 - \dfrac{d}{f}\right) + 1 = 0$$

Let us introduce the abbreviation $\alpha = 1 - d/2f$. The roots of the above quadratic equation are then found to be

$$\lambda = \alpha \pm \sqrt{\alpha^2 - 1} = e^{\pm \phi} \qquad |\alpha| > 1 \qquad (10.28)$$

or

$$\lambda = \alpha \pm i \sqrt{1 - \alpha^2} = e^{\pm i\phi} \qquad |\alpha| < 1 \qquad (10.29)$$

in which ϕ is a real number.

Now suppose that a given ray vector is an eigenvector of the system, and that this ray passes through a number N of repeated reflections or refractions in the optical system. The final output ray vector is then

$$\begin{bmatrix} \rho_N \\ \theta_N \end{bmatrix} = \lambda^N \begin{bmatrix} \rho_1 \\ \theta_1 \end{bmatrix} \tag{10.30}$$

It is clear that the ray trajectory will remain *stable* or stay close to the axis of the optical system if the condition expressed by Equation (10.29) holds, because in that case $\lambda^N = e^{\pm iN\phi}$ and $|\lambda^N| = 1$. In the *unstable* case of Equation (10.28) the trajectory will diverge. In terms of the optical parameters the stability criterion $|\alpha| < 1$ gives

$$0 < d < 4f \tag{10.31}$$

or in the case of the optical resonator

$$0 < d < 2r \tag{10.32}$$

The focal length must be positive (converging lens or mirror) and the separation must be less than four times the focal length, or twice the radius of curvature of the resonator mirrors. In the *confocal* configuration, $d = 2f = r$. This satisfies the stability requirement and is often the configuration employed for laser resonators.

For the unsymmetrical optical resonator or periodic waveguide consisting of two mirrors of focal length f_1 and f_2 separated by a distance d, or by alternate lenses of focal length f_1 and f_2 all separated by equal distances d, a similar procedure to that outlined above yields the following stability criterion:

$$0 < \alpha_1 \alpha_2 < 1 \tag{10.33}$$

in which

$$\alpha_1 = 1 - \frac{d}{2f_1} \qquad \alpha_2 = 1 - \frac{d}{2f_2} \tag{10.34}$$

The calculation is left as an exercise. See problems 10.6 and 10.7.

PROBLEMS

10.1   Derive Equation (10.7) by ray tracing.

10.2   Determine the focal length and the positions of the principal planes for a lens consisting of a glass sphere of radius r and index n.

10.3   Prove Equation (10.13).

10.4   Use the ray matrix method to prove Equation (10.7).

10.5 Use the ray matrix method to derive the focal length of a thick lens, Equation (10.11).

10.6 Derive the stability condition for the unsymmetrical optical resonator, Equation (10.33).

10.7 Illustrate the stability condition $0 < \alpha_1\alpha_2 < 1$ graphically by plotting α_1 versus α_2 as rectangular coordinates. To do this, plot the two branches of the hyperbola $\alpha_1\alpha_2 = 1$. Show that the stability condition is satisfied in the first and third quadrants in the region bounded by the coordinate axes and the corresponding branch of the hyberbola. Note that for an optical resonator, the origin $\alpha_1 = 0 = 1 - d/r_1$, $\alpha_2 = 0 = 1 - d/r_2$ represents the confocal resonator: $d = r_1 = r_2$. Similarly, the point $\alpha_2 = \alpha_1 = 1$ represents the plane-parallel resonator. Which point corresponds to the hemispherical resonator?

APPENDIX I

Relativistic Optics

I.1 The Michelson-Morley Experiment

The famous Michelson-Morley experiment, performed in 1887, was designed to measure the absolute velocity of the earth's motion in space by means of light waves.

A diagram of the optical arrangement is shown in Figure I.1. The

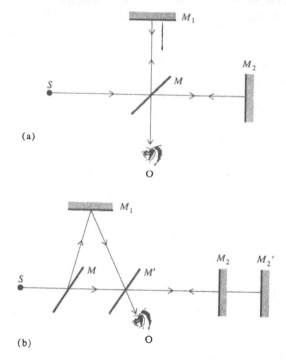

(a)

(b)

Figure I.1. Simplified diagram of the Michelson-Morley experiment.

apparatus is essentially an optical interferometer. A beam of light from a source S is split into two beams by a half-silvered mirror M. One beam is reflected to a mirror M_1, which in turn reflects the light directly back to M. The other beam is transmitted directly to the mirror M_2, which also reflects the light back to M. The two partial beams then unite at M, part of the combined light going to an observer O who sees an interference pattern of bright and dark fringes. The interference pattern can be made to shift by one fringe by displacing either of the two mirrors M_1 or M_2 a distance of $\frac{1}{4}$ wavelength.

If mirrors M_1 and M_2 are both located at precisely the same distance from M and if the apparatus does not move during the time that light is reflected back and forth, then the two waves return to M at the same phase so that a bright fringe is seen at O. Suppose, however, that the whole apparatus is moving in the direction of the initial beam SM. The paths of the beams will then be as shown by the directed lines in the figure. The times taken by the two partial waves in their respective journeys are no longer the same if it is assumed that light travels with a constant speed c in some medium. The situation is analogous to the case of two swimmers in a stream, one swimmer going upstream and back, the other going across the stream and returning.

To analyze the situation quantitatively, let us suppose that the speed of the apparatus through the medium is u. Then the wave moving toward M_2 travels with a speed $c - u$ relative to the apparatus. On its return this wave travels with relative speed $c + u$. The total time for the round trip is therefore

$$t_2 = \frac{d}{c - u} + \frac{d}{c + u} = \frac{2cd}{c^2 - u^2}$$

in which d is the distance OM_2. On the other hand, the wave reflected by M_1 travels along the path $MM_1'O$, as shown. If we call t_1 the total time for the round trip in this case, then the distance MM_1' is equal to $\sqrt{d^2 + (\frac{1}{4})u^2 t_1{}^2}$. Thus

$$t_1 = \left(\frac{2}{c}\right) \sqrt{d^2 + \tfrac{1}{4} u^2 t_1{}^2}$$

Solving for t_1, we get

$$t_1 = \frac{2d}{\sqrt{c^2 - u^2}}$$

The time difference Δt between the two paths is accordingly

$$\Delta t = t_2 - t_1 = 2d \left(\frac{c}{c^2 - u^2} - \frac{1}{\sqrt{c^2 - u^2}}\right) = \frac{du^2}{c^3} + \cdots$$

This corresponds to a phase difference

$$\Delta\phi = \omega \, \Delta t = \frac{2\pi c}{\lambda} \, \Delta t \approx \frac{2\pi d}{\lambda} \frac{u^2}{c^2}$$

where λ is the wavelength of the light.

In their experiment, Michelson and Morley obtained an effective distance d of 10 m by multiple reflections as indicated in Figure I.2.

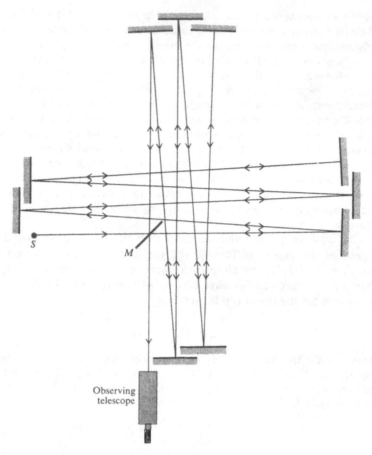

Figure I.2.   Actual light path in the Michelson-Morley experiment.

The experiment was performed by floating the entire apparatus in a pool of mercury and observing the fringes as the apparatus was rotated through an angle of 90 degrees. This would cause either of the two beams to be alternately parallel or perpendicular to the earth's motion. In its orbital motion around the sun, the earth's speed is about $10^{-4}\,c$. The expected shift with yellow light, 5900 Å, was about one third of a fringe. *Actually there was no observable shift at all.* This negative result came as a surprise to the scientific world. It was in contradiction to the (then) accepted idea concerning electromagnetic radiation, namely, that such radiation must have a medium for its transmission through space. This medium, called the *ether*, was supposed to be an all-pervading substance, and numerous calculations concerning its properties had been carried out, including some by Maxwell.

The Michelson-Morley experiment has been repeated many times by different observers with essentially the same negative results. There have been some reports of measurable fringe shifts, but none anywhere near as large as should be predicted by the orbital speed of the earth. This is actually a minimum speed since the speed of the whole solar system, due to rotation of our galaxy, is about ten times the earth's orbital speed.

The idea of the ether had been so widely accepted that it was many years before it was finally abandoned. In fact two physicists, Fitzgerald and Lorentz, proposed to explain the null result of the Michelson-Morley experiment by suggesting that a body *contracts* in the direction of its motion through the ether in precisely the ratio $\sqrt{1 - u^2/c^2}$. This amount of shortening, known as the *Fitzgerald-Lorentz contraction*, would just equalize the two light paths so that there would be no fringe shift. Now such an ad hoc explanation of the experiment is not very satisfactory, for the contraction is not capable of direct observation. Any attempt to measure it would fail, since the measuring apparatus contracts along with the object to be measured.

I.2   Einstein's Postulates of Special Relativity

In 1905 Albert Einstein formulated his special theory of relativity. This theory is based on two fundamental postulates:

(1) *All physical laws have the same form in all inertial coordinate systems*
(2) *The speed of electromagnetic radiation in the vacuum is the same in all inertial systems*

The first postulate is a statement concerning physical laws in general and is an extension of Newtonian relativity. It can be shown that Maxwell's equations obey this postulate; that is, the equations have the same general form in any inertial coordinate system. The proof is given in almost any textbook on relativity [32].

The second postulate is more specific. It is the one that is of immediate application to our study of optics. It says that any measurement of the speed of light must always yield the same result, even if the source of light is in motion relative to the observer, or if the observer is moving relative to the source. This postulate immediately explains the null result of the Michelson-Morley experiment, for it implies that the speed of propagation of each beam in the experimental arrangement is always c, whether the apparatus is moving or not. Hence there is no phase change and no fringe shift.[1]

I.3 Relativistic Effects in Optics

According to the second postulate of the special theory of relativity, *the speed of light in vacuum is the same for any observer, regardless of the motion of the source relative to him or of his motion relative to the source.* To examine the consequences of this postulate, let us consider two observers moving with constant relative speed u. We shall designate the coordinate systems of our two observers by $Oxyz$ and $O'x'y'z'$, respectively. For simplicity, we shall assume that the respective axes Ox, $O'x'$, and so forth, are parallel, and that the relative motion is in the xx' direction (Figure I.3).

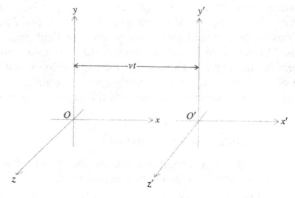

Figure I.3. Coordinate systems of two observers moving with constant relative speed.

Suppose that the two origins O and O' are coincident at time $t = 0$. Then the distance OO' is equal to ut, and the equations of transformation, according to classical or Newtonian kinematics, are

$$x = x' + ut'$$

$$y = y'$$

$$z = z'$$

$$t = t'$$

---

[1] It should be noted that Einstein did not formulate the theory of relativity in order to account for the Michelson-Morley experiment. Rather, the Michelson-Morley experiment has merely been cited as one experiment that tends to confirm the second postulate. Other, more recent experiments have been performed to verify the constancy of the velocity of light when the source and the observer are in relative motion. An excellent discussion and review of these is given by J. G. Fox, *Amer. J. Phys.*, **33**, 1 (1965).

The equation $t = t'$ expresses the assumed equality of the time scales of the two observers. They are using identical clocks. The above equations of transformation clearly contradict the second postulate, because from them we obtain $dx/dt = dx'/dt' + u$, so that anything moving with the speed of light c in, say, the primed system, moves with the speed $c + u$ in the unprimed system.

In order to find a coordinate transformation that agrees with the second postulate of relativity, let us consider the wave equation

$$\frac{\partial^2 U}{\partial x^2} - \frac{1}{c^2}\frac{\partial^2 U}{\partial t^2} = 0$$

This is the differential equation of a light wave propagating with velocity c in the x direction. The requirement of the second postulate is that the equation remains invariant when referred to the primed coordinate system. That is,

$$\frac{\partial^2 U}{\partial x'^2} - \frac{1}{c^2}\frac{\partial^2 U}{\partial t'^2} = 0$$

This will be the case if

$$\frac{\partial^2 U}{\partial x^2} - \frac{1}{c^2}\frac{\partial^2 U}{\partial t^2} = \frac{\partial^2 U}{\partial x'^2} - \frac{1}{c^2}\frac{\partial^2 U}{\partial t'^2} \qquad \text{(I.1)}$$

Now it turns out that a general linear transformation[2] of the form

$$x = a_{11}x' + a_{12}t'$$
$$t = a_{21}x' + a_{22}t'$$

with the proper choice of constants will make the wave equation invariant. By substituting the above transformation in (I.1) and requiring that the equation be an identity, we obtain three equations to solve for the coefficients a_{11}, a_{12}, and so forth. We also need the subsidiary condition that $x = 0$ transforms to $x' = -ut'$, so that $a_{12} - ua_{11}$. The result is the famous Lorentz transformation,

$$x = \gamma(x' + ut')$$
$$y = y' \qquad \text{(I.2)}$$
$$z = z'$$
$$t = \gamma\left(t' + \frac{ux'}{c^2}\right)$$

where

$$\gamma = \frac{1}{\sqrt{1 - u^2/c^2}}$$

[2] A nonlinear transformation would be unrealistic because uniform motion in one coordinate system would appear as accelerated motion in the other.

It is assumed that the kinematic consequences of the Lorentz transformation, for example, length contraction and time dilatation, are already familiar to the reader [32].

The Relativistic Doppler Formula   Let us now consider a plane electromagnetic wave whose space–time dependence is of the form exp $i(kx - \omega t)$ when represented in the unprimed coordinate system. The observer in this system sees a wave of angular frequency $\omega = ck$ moving in the x direction. By applying the Lorentz transformation we find that the observer in the primed coordinate system sees the space–time dependence of this same wave as

$$\exp i\left[k\gamma(x' + ut') - \omega\gamma\left(t' + \frac{ux'}{c^2}\right)\right]$$
$$= \exp i\left[\left(k\gamma - \frac{\omega\gamma u}{c^2}\right) x' - (\omega\gamma - k\gamma u)\, t'\right] \tag{I.3}$$

This must be identical with the expression

$$\exp i(k'x' - \omega't')$$

Hence

$$\omega' = \omega\gamma\left(1 - \frac{ku}{\omega}\right) = \omega\gamma\left(1 - \frac{u}{c}\right) \tag{I.4}$$

Also, since $\omega = 2\pi\nu$ and $\gamma = [1 - (u^2/c^2)]^{-1/2}$, we can write

$$\nu' = \nu\frac{1 - u/c}{\sqrt{1 - u^2/c^2}} = \nu\frac{\sqrt{1 - u/c}}{\sqrt{1 + u/c}} = \nu\left(1 - \frac{u}{c} + \frac{1}{2}\frac{u^2}{c^2} - \cdots\right) \tag{I.5}$$

This is the relativistic Doppler formula. The series expansion shows that the relativistic Doppler shift differs from the nonrelativistic values only in the second- and higher-order terms, and therefore this difference becomes important only for large velocities. The relativistic formula has been verified by experiments with high-speed hydrogen atoms in a specially designed discharge tube shown in Figure I.4 [20].

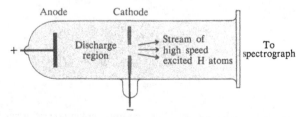

Figure I.4.   Discharge tube used to observe the relativistic Doppler effect.

The Transverse Doppler Shift   Let us next suppose that we have a plane wave traveling in the negative y direction in the unprimed system. The space–time dependence of this wave is $\exp i(ky + \omega t)$. By applying the Lorentz transformation we find that to an observer in the primed system moving with speed v in the x direction, the space–time dependence of the wave is

$$\exp i \left[ky' + \omega \left(\gamma t' + \frac{ux'\gamma}{c^2} \right) \right] = \exp i \left[\frac{\omega u \gamma x'}{c^2} + ky' + \omega \gamma t' \right]$$

Since this must be the same as

$$\exp i \, (k_x'x' + k_y'u' + \omega't')$$

then, for the coefficient of t' we have

$$\omega' = \omega\gamma$$

or, equivalently,

$$\nu = \nu' \sqrt{1 - \frac{u^2}{c^2}} = \nu' \left(1 - \frac{u^2}{2c^2} + \cdots \right) \tag{I.6}$$

This is the formula for the *transverse Doppler shift,* giving the frequency change when the relative motion is at right angles to the direction of observation. The transverse Doppler shift is a second-order effect and is therefore very difficult to measure. It has been verified by using the Mossbauer effect with gamma radiation from radioactive atoms [11].

The Aberration of Starlight   Another consequence of the relativistic transformation of a plane wave moving in the y-direction is the appearance of x' in the wave function. This implies that the wave vector $\mathbf{k}'$ has a component in the x' direction and, consequently, the direction of propagation is not exactly the same as the direction of the y' axis. The angle α of the inclination to the y' axis is given by $\tan \alpha = k_{x'}/k_{y'}$. Hence

$$\tan \alpha = \frac{\omega \gamma u/c^2}{k} = \frac{u}{c} \gamma = \frac{u}{c\sqrt{1 - u^2/c^2}} \tag{I.7}$$

This effect is known as *aberration of light.* It was first observed experimentally by the English astronomer Bradley in 1727. Bradley found an apparent shift in the positions of stars that was greatest for those stars whose line of sight was at right angles to the earth's orbital velocity around the sun. The maximum value of this stellar aberration is about 20 s of arc. Bradley's explanation is illustrated in Figure I.5, which shows the change in apparent direction due to the observer's velocity v. The situation is similar to that of a person running through

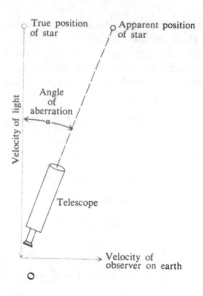

Figure I.5.   The aberration of starlight.

falling rain. If the rain is falling straight down, its velocity relative to the person is not vertical but has a horizontal component equal to the forward speed of the person. From the figure we have $\tan \alpha = u/c$. This simple formula differs from the relativistic formula (I.7) by the factor γ. In the case of the earth, however, u/c is of the order of 10^{-4}, so the difference is entirely negligible.

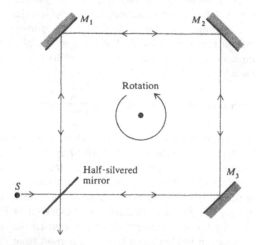

Figure I.6.   Diagram of Sagnac's experiment.

It is interesting to note that the nonrelativistic transformation gives zero aberration for plane waves, hence aberration is a relativistic effect in this context. The simple explanation is valid if light is considered to be a hail of photons, however.

I.4 The Experiments of Sagnac and of Michelson and Gale to Detect Rotation

In 1911 the French physicist G. Sagnac performed an interesting experiment designed to detect rotation by means of light beams. His ex-

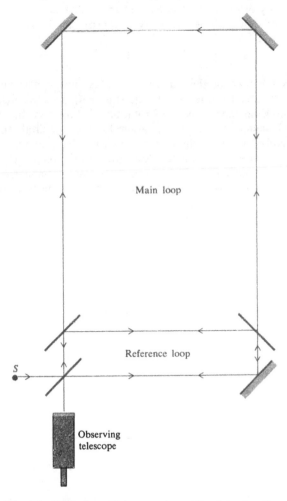

Figure I.7. The Michelson-Gale experiment for detecting the absolute rotation of the earth.

periment is illustrated in Figure I.6. A beam of light from a source S is divided into two beams by means of a half-silvered mirror M. The two beams are caused to traverse opposing paths around a circuit formed by mirrors M_1, M_2, and M_3 as shown. The beams recombine at M and are reflected into an observing telescope in which interference fringes are seen.

The apparatus is mounted on a rigid support that can be rotated about a vertical axis. The rotation causes a difference in the time required for the clockwise and counterclockwise beams to traverse the circuit. The result is a fringe shift that is proportional to the angular velocity of rotation. It is easy to show that the effective path difference Δs for the two beams is given approximately by

$$\Delta s = \frac{4A}{c}\,\Omega$$

where A is the area of the circuit and Ω is the angular velocity.

Sagnac was able to observe a fringe shift with a square light path about 1 m on a side and a speed of rotation of 120 r/min. In order to detect small angular velocities, a larger loop is required. In 1925 Michelson and Gale set up the experiment with a large path, 2/5 mile by 1/5 mile (Figure I.7). With this loop they were able to detect the expected fringe shift due to rotation of the earth. A smaller loop inside the larger one was used to provide a set of reference fringes.

REFERENCES

1. Beran, M. J., and G. B. Parrent, Jr., *Theory of Partial Coherence*. Englewood Cliffs, N. J.: Prentice-Hall, 1964.
2. Besancon, R. M., ed., *The Encyclopedia of Physics*. New York: Reinhold, 1966.
3. Bloembergen, N., *Nonlinear Optics*. New York: W. A. Benjamin, 1965.
4. Bloom, A. L., W. E. Bell, and F. O. Lopez, *Phys. Rev.*, **135**, A578 (1964).
5. Born, M., and E. Wolf, *Principles of Optics*. New York: Macmillan, 1964.
6. Boyd, G. D., and J. P. Gordon, *Bell System Tech. J.*, **40**, 489 (1961).
7. Candler, C., *Modern Interferometers*. London: Hilger and Watts, 1951.
8. Fowles, G. R., and W. T. Silfvast, *J. Quantum Electronics*, **QE-1**, 131 (1965).
9. Fox, A. G., and T. Li, *Bell System Tech. J.*, **40**, 453 (1961).
10. Francon, M., *Modern Applications of Physical Optics*. New York: Wiley, 1963.
11. Frauenfelder, H., *The Mössbauer Effect*. New York: W. A. Benjamin, 1962.
12. Gabor, D., *Nature*, **161**, 777 (1948).
13. Gordon, J. P., H. Z. Zeiger, and C. H. Townes, *Phys. Rev.*, **95**, 282 (1954).
14. Gray, D. E. ed., *American Institute of Physics Handbook*. New York: McGraw-Hill, 1957.
15. Hanbury-Brown, R., and R. Q. Twiss, *Proc. Roy. Soc. (London)*, **A243**, 291 (1957).
16. Harnwell, G. P., *Principles of Electricity and Electromagnetism*. New York: McGraw-Hill, 1938.
17. Harrison, G. R., R. C. Lord, and J. R. Loofbourow, *Practical Spectroscopy*. Englewood Cliffs, N. J.: Prentice-Hall, 1948.
18. Herzberg, G., *Atomic Spectra and Atomic Structure*. New York: Dover, 1950.
19. ———, *Molecular Spectra and Molecular Structure*. Princeton, N. J.: Van Nostrand, 1950.
20. Ives, H. E., and G. R. Stilwell, *J. Opt. Soc. Am.*, **31**, 369 (1941).
21. Javan, A., W. R. Bennet, Jr., and D. R. Herriott, *Phys. Rev. Letters*, **6**, 106 (1961).
22. Jensen, R. C., and G. R. Fowles, *Proc. IEEE*, **52**, 1350 (1964).
23. King, G. W., *Spectroscopy and Molecular Structure*. New York: Holt, Rinehart and Winston, 1964.

24. Kuhn, H. G., *Atomic Spectra*. New York: Academic Press, 1962.
25. Lengyel, B. A., *Introduction to Laser Physics*. New York: Wiley, 1966.
26. Lilley, A. E. *et al.*, *Nature*, **209**, 468 (1966).
27. Mathews, J., and R. L. Walker, *Mathematical Methods of Physics*. New York: W. A. Benjamin, 1964.
28. Nicols, E. F., and G. F. Hull, *Phys. Rev.*, **13**, 307 (1901).
29. Pearson, J. M., *A Theory of Waves*. Boston: Allyn and Bacon, 1966.
30. Prather, J. L., *Atomic Energy Levels in Crystals*. Washington, D. C.: Nat. Bur. Stand. Monograph 19, U. S. Govt. Printing Office, 1961.
31. Present, R. D., *Kinetic Theory of Gases*. New York: McGraw-Hill, 1958.
32. Rindler, W., *Special Relativity*. London: Olives & Boyd, 1960.
33. Rosa, E. B., and N. E. Dorsey, *A New Determination of the Ratio of the Electromagnetic to the Electrostatic Unit of Electricity*. Washington, D.C.: U.S. Bureau of Standards, Reprint No. 65, 1907.
34. Rossi, B., *Optics*. Reading, Mass.: Addison-Wesley, 1957.
35. Sawyer, R. A., *Experimental Spectroscopy*. Englewood Cliffs, N. J.: Prentice-Hall, 1944.
36. Shurcliff, W. A., and S. S. Ballard, *Polarized Light*. Princeton, N. J.: Van Nostrand, 1964.
37. Silfvast, W. T., G. R. Fowles, and B. H. Hopkins, *Appl. Phys. Letters*, **8**, 318 (1966).
38. Stroke, G. W., *An Introduction to Coherent Optics and Holography*, 2nd ed. New York: Academic Press, 1969.
39. West, C. D., and R. C. Jones, *J. Opt. Soc. Amer.*, **41**, 975 (1951).
40. White, H. E., *Introduction to Atomic Spectra*. New York: McGraw-Hill, 1934.
41. Williams, W. E., *Applications of Interferometry*, 4th ed. New York: Wiley, 1950.
42. Zernike, F. and J. E. Midwinter, *Applied Nonlinear Optics*. New York: Wiley, 1973.

ANSWERS TO SELECTED
ODD-NUMBERED PROBLEMS

Chapter 1

1.1   (a) $U = U_0 \cos 2\pi \left(\dfrac{z}{\lambda} - t\right)$, (c) $U = U_0 \cos \omega \left(\dfrac{z}{u} - t\right)$

1.5 1.3×10^7 m^{-1}

1.7 $u = c/1.62$ $u_g = c/1.86$

1.11 1.54×10^9 Hz, 1.84×10^{-3} nm

Chapter 2

2.3 1.27×10^{18} W/m^2, 3.09×10^{10} V/m

2.7   (a) $\begin{bmatrix} 1/2 \\ 1/2 \end{bmatrix}$, (b) $\begin{bmatrix} 1 \\ 2 \end{bmatrix}$, (c) $\begin{bmatrix} 1 \\ -i \end{bmatrix}$ (d) $\begin{bmatrix} \sqrt{2} \\ 1+i \end{bmatrix}$

2.9   The real fields in the x- and y-directions are $A \cos \omega t$ and $B \cos (\omega t + \Delta)$. Elimination of the parameter t between these two expressions gives an equation of an ellipse. Transform to a rotated coordinate system x', y' and choose the angle of rotation such that the term in $x'y'$ vanishes.

2.13 $\lambda = 0$ gives the eigenvector $A \begin{bmatrix} 1 \\ -1 \end{bmatrix}$, and $\lambda = 2$ gives $A \begin{bmatrix} 1 \\ 1 \end{bmatrix}$ where A is an arbitrary constant.

2.15   53 degrees (water), 68 degrees (diamond)

2.17   55 degrees

2.19   (a) 0.00033 mm, (b) 10^{-2570}

Chapter 3

3.1 $I/I_0 = 3 + 4 \cos \theta + 2 \cos 2\theta$ where $\theta = kyh/x$

3.3   The equivalent optical path difference is $d(n - 1) = 0.025$ mm where d is the glass thickness. Hence the lateral shift is given by the expression $\dfrac{xd\,(n - 1)}{h} = 5$ cm.

3.5 $\dfrac{\lambda(D + D')}{\alpha D(n - 1)}$

3.7   0.036 mm, 1.2×10^{-13} s

3.9   0.08 mm

3.11   16 cm

Chapter 4

4.1 $\mathscr{T}_{max} = 0.25$, $\mathscr{T}_{min} = 0.0007$

4.3   480 KHz, 6.4×10^{-7} nm

4.5 $T = \left[1 + \left(\dfrac{n^2 - 1}{2n}\right)^2 \sin^2 kd\right]^{-1}$ where $k = 2\pi/\lambda = 2\pi n/\lambda_0$

4.7 $R = 0.8$ percent

Chapter 5

5.1   (a) Fresnel, (b) Fraunhofer

5.3   507 nm

5.7   58 cm

5.11   about 0.045

5.13   about 105 cm (Good optical gratings this wide are not made, hence resolution of laser modes requires a Fabry-Perot interferometer.)

5.15 $I/I' = 4$ (Open part of aperture contains exactly 3 Fresnel zones.)

5.21 $g'(y') = 2 \ \text{Si}[2\pi b/(b - 2y')] + 2 \ \text{Si}[2\pi b/(b + 2y')]$ where Si is the *sine integral*, $\text{Si}(y) = \displaystyle\int_0^y (\sin u)/u \ du$.

Chapter 6

6.5 $n = 9 \pm \sqrt{68} = 0.75, \ 17.25$ (The smaller root is the more physically realistic.)

6.13   40.25 degrees

6.15   0.012 degrees

6.17   1.3 degrees

Chapter 7

7.1   (a) 3.4×10^5, (b) 4×10^{13}

7.3 2.4×10^{14}

7.5 $3.2 \times 10^7 V T^3$

7.7 $I_\lambda = \dfrac{2\pi hc}{\lambda^5 (e^{hc/\lambda kT} - 1)}$

7.9   (a) 3.2×10^{15}, (b) 4.7×10^9

7.11   35,600°K

Chapter 8

8.3   6.6 GHz
8.5   (a) 0.33, (b) $4A_H$ (approx.)
8.9   about 3 Terahertz (THz)

Chapter 9

9.1   2.1 eV
9.3 1.6×10^{-25}
9.5 $9 \times 10^{15}/cm^3$
9.7   1.5 mm

Index

323

A CATALOG OF SELECTED
DOVER BOOKS
IN SCIENCE AND MATHEMATICS

Astronomy

CHARIOTS FOR APOLLO: The NASA History of Manned Lunar Spacecraft to 1969, Courtney G. Brooks, James M. Grimwood, and Loyd S. Swenson, Jr. This illustrated history by a trio of experts is the definitive reference on the Apollo spacecraft and lunar modules. It traces the vehicles' design, development, and operation in space. More than 100 photographs and illustrations. 576pp. 6 3/4 x 9 1/4. 0-486-46756-2

EXPLORING THE MOON THROUGH BINOCULARS AND SMALL TELESCOPES, Ernest H. Cherrington, Jr. Informative, profusely illustrated guide to locating and identifying craters, rills, seas, mountains, other lunar features. Newly revised and updated with special section of new photos. Over 100 photos and diagrams. 240pp. 8 1/4 x 11. 0-486-24491-1

WHERE NO MAN HAS GONE BEFORE: A History of NASA's Apollo Lunar Expeditions, William David Compton. Introduction by Paul Dickson. This official NASA history traces behind-the-scenes conflicts and cooperation between scientists and engineers. The first half concerns preparations for the Moon landings, and the second half documents the flights that followed Apollo 11. 1989 edition. 432pp. 7 x 10. 0-486-47888-2

APOLLO EXPEDITIONS TO THE MOON: The NASA History, Edited by Edgar M. Cortright. Official NASA publication marks the 40th anniversary of the first lunar landing and features essays by project participants recalling engineering and administrative challenges. Accessible, jargon-free accounts, highlighted by numerous illustrations. 336pp. 8 3/8 x 10 7/8. 0-486-47175-6

ON MARS: Exploration of the Red Planet, 1958-1978--The NASA History, Edward Clinton Ezell and Linda Neuman Ezell. NASA's official history chronicles the start of our explorations of our planetary neighbor. It recounts cooperation among government, industry, and academia, and it features dozens of photos from Viking cameras. 560pp. 6 3/4 x 9 1/4. 0-486-46757-0

ARISTARCHUS OF SAMOS: The Ancient Copernicus, Sir Thomas Heath. Heath's history of astronomy ranges from Homer and Hesiod to Aristarchus and includes quotes from numerous thinkers, compilers, and scholasticists from Thales and Anaximander through Pythagoras, Plato, Aristotle, and Heraclides. 34 figures. 448pp. 5 3/8 x 8 1/2. 0-486-43886-4

AN INTRODUCTION TO CELESTIAL MECHANICS, Forest Ray Moulton. Classic text still unsurpassed in presentation of fundamental principles. Covers rectilinear motion, central forces, problems of two and three bodies, much more. Includes over 200 problems, some with answers. 437pp. 5 3/8 x 8 1/2. 0-486-64687-4

BEYOND THE ATMOSPHERE: Early Years of Space Science, Homer E. Newell. This exciting survey is the work of a top NASA administrator who chronicles technological advances, the relationship of space science to general science, and the space program's social, political, and economic contexts. 528pp. 6 3/4 x 9 1/4. 0-486-47464-X

STAR LORE: Myths, Legends, and Facts, William Tyler Olcott. Captivating retellings of the origins and histories of ancient star groups include Pegasus, Ursa Major, Pleiades, signs of the zodiac, and other constellations. "Classic." – *Sky & Telescope.* 58 illustrations. 544pp. 5 3/8 x 8 1/2. 0-486-43581-4

A COMPLETE MANUAL OF AMATEUR ASTRONOMY: Tools and Techniques for Astronomical Observations, P. Clay Sherrod with Thomas L. Koed. Concise, highly readable book discusses the selection, set-up, and maintenance of a telescope; amateur studies of the sun; lunar topography and occultations; and more. 124 figures. 26 halftones. 37 tables. 335pp. 6 1/2 x 9 1/4. 0-486-42820-6

Browse over 9,000 books at www.doverpublications.com

Chemistry

MOLECULAR COLLISION THEORY, M. S. Child. This high-level monograph offers an analytical treatment of classical scattering by a central force, quantum scattering by a central force, elastic scattering phase shifts, and semi-classical elastic scattering. 1974 edition. 310pp. 5 3/8 x 8 1/2.   0-486-69437-2

HANDBOOK OF COMPUTATIONAL QUANTUM CHEMISTRY, David B. Cook. This comprehensive text provides upper-level undergraduates and graduate students with an accessible introduction to the implementation of quantum ideas in molecular modeling, exploring practical applications alongside theoretical explanations. 1998 edition. 832pp. 5 3/8 x 8 1/2.   0-486-44307-8

RADIOACTIVE SUBSTANCES, Marie Curie. The celebrated scientist's thesis, which directly preceded her 1903 Nobel Prize, discusses establishing atomic character of radioactivity; extraction from pitchblende of polonium and radium; isolation of pure radium chloride; more. 96pp. 5 3/8 x 8 1/2.   0-486-42550-9

CHEMICAL MAGIC, Leonard A. Ford. Classic guide provides intriguing entertainment while elucidating sound scientific principles, with more than 100 unusual stunts: cold fire, dust explosions, a nylon rope trick, a disappearing beaker, much more. 128pp. 5 3/8 x 8 1/2.   0-486-67628-5

ALCHEMY, E. J. Holmyard. Classic study by noted authority covers 2,000 years of alchemical history: religious, mystical overtones; apparatus; signs, symbols, and secret terms; advent of scientific method, much more. Illustrated. 320pp. 5 3/8 x 8 1/2.
0-486-26298-7

CHEMICAL KINETICS AND REACTION DYNAMICS, Paul L. Houston. This text teaches the principles underlying modern chemical kinetics in a clear, direct fashion, using several examples to enhance basic understanding. Solutions to selected problems. 2001 edition. 352pp. 8 3/8 x 11.   0-486-45334-0

PROBLEMS AND SOLUTIONS IN QUANTUM CHEMISTRY AND PHYSICS, Charles S. Johnson and Lee G. Pedersen. Unusually varied problems, with detailed solutions, cover of quantum mechanics, wave mechanics, angular momentum, molecular spectroscopy, scattering theory, more. 280 problems, plus 139 supplementary exercises. 430pp. 6 1/2 x 9 1/4.   0-486-65236-X

ELEMENTS OF CHEMISTRY, Antoine Lavoisier. Monumental classic by the founder of modern chemistry features first explicit statement of law of conservation of matter in chemical change, and more. Facsimile reprint of original (1790) Kerr translation. 539pp. 5 3/8 x 8 1/2.   0-486-64624-6

MAGNETISM AND TRANSITION METAL COMPLEXES, F. E. Mabbs and D. J. Machin. A detailed view of the calculation methods involved in the magnetic properties of transition metal complexes, this volume offers sufficient background for original work in the field. 1973 edition. 240pp. 5 3/8 x 8 1/2.   0-486-46284-6

GENERAL CHEMISTRY, Linus Pauling. Revised third edition of classic first-year text by Nobel laureate. Atomic and molecular structure, quantum mechanics, statistical mechanics, thermodynamics correlated with descriptive chemistry. Problems. 992pp. 5 3/8 x 8 1/2.   0-486-65622-5

ELECTROLYTE SOLUTIONS: Second Revised Edition, R. A. Robinson and R. H. Stokes. Classic text deals primarily with measurement, interpretation of conductance, chemical potential, and diffusion in electrolyte solutions. Detailed theoretical interpretations, plus extensive tables of thermodynamic and transport properties. 1970 edition. 590pp. 5 3/8 x 8 1/2.   0-486-42225-9

Engineering

FUNDAMENTALS OF ASTRODYNAMICS, Roger R. Bate, Donald D. Mueller, and Jerry E. White. Teaching text developed by U.S. Air Force Academy develops the basic two-body and n-body equations of motion; orbit determination; classical orbital elements, coordinate transformations; differential correction; more. 1971 edition. 455pp. 5 3/8 x 8 1/2.   0-486-60061-0

INTRODUCTION TO CONTINUUM MECHANICS FOR ENGINEERS: Revised Edition, Ray M. Bowen. This self-contained text introduces classical continuum models within a modern framework. Its numerous exercises illustrate the governing principles, linearizations, and other approximations that constitute classical continuum models. 2007 edition. 320pp. 6 1/8 x 9 1/4.   0-486-47460-7

ENGINEERING MECHANICS FOR STRUCTURES, Louis L. Bucciarelli. This text explores the mechanics of solids and statics as well as the strength of materials and elasticity theory. Its many design exercises encourage creative initiative and systems thinking. 2009 edition. 320pp. 6 1/8 x 9 1/4.   0-486-46855-0

FEEDBACK CONTROL THEORY, John C. Doyle, Bruce A. Francis and Allen R. Tannenbaum. This excellent introduction to feedback control system design offers a theoretical approach that captures the essential issues and can be applied to a wide range of practical problems. 1992 edition. 224pp. 6 1/2 x 9 1/4.   0-486-46933-6

THE FORCES OF MATTER, Michael Faraday. These lectures by a famous inventor offer an easy-to-understand introduction to the interactions of the universe's physical forces. Six essays explore gravitation, cohesion, chemical affinity, heat, magnetism, and electricity. 1993 edition. 96pp. 5 3/8 x 8 1/2.   0-486-47482-8

DYNAMICS, Lawrence E. Goodman and William H. Warner. Beginning engineering text introduces calculus of vectors, particle motion, dynamics of particle systems and plane rigid bodies, technical applications in plane motions, and more. Exercises and answers in every chapter. 619pp. 5 3/8 x 8 1/2.   0-486-42006-X

ADAPTIVE FILTERING PREDICTION AND CONTROL, Graham C. Goodwin and Kwai Sang Sin. This unified survey focuses on linear discrete-time systems and explores natural extensions to nonlinear systems. It emphasizes discrete-time systems, summarizing theoretical and practical aspects of a large class of adaptive algorithms. 1984 edition. 560pp. 6 1/2 x 9 1/4.   0-486-46932-8

INDUCTANCE CALCULATIONS, Frederick W. Grover. This authoritative reference enables the design of virtually every type of inductor. It features a single simple formula for each type of inductor, together with tables containing essential numerical factors. 1946 edition. 304pp. 5 3/8 x 8 1/2.   0-486-47440-2

THERMODYNAMICS: Foundations and Applications, Elias P. Gyftopoulos and Gian Paolo Beretta. Designed by two MIT professors, this authoritative text discusses basic concepts and applications in detail, emphasizing generality, definitions, and logical consistency. More than 300 solved problems cover realistic energy systems and processes. 800pp. 6 1/8 x 9 1/4.   0-486-43932-1

THE FINITE ELEMENT METHOD: Linear Static and Dynamic Finite Element Analysis, Thomas J. R. Hughes. Text for students without in-depth mathematical training, this text includes a comprehensive presentation and analysis of algorithms of time-dependent phenomena plus beam, plate, and shell theories. Solution guide available upon request. 672pp. 6 1/2 x 9 1/4.   0-486-41181-8

Browse over 9,000 books at www.doverpublications.com

HELICOPTER THEORY, Wayne Johnson. Monumental engineering text covers vertical flight, forward flight, performance, mathematics of rotating systems, rotary wing dynamics and aerodynamics, aeroelasticity, stability and control, stall, noise, and more. 189 illustrations. 1980 edition. 1089pp. 5 5/8 x 8 1/4.   0-486-68230-7

MATHEMATICAL HANDBOOK FOR SCIENTISTS AND ENGINEERS: Definitions, Theorems, and Formulas for Reference and Review, Granino A. Korn and Theresa M. Korn. Convenient access to information from every area of mathematics: Fourier transforms, Z transforms, linear and nonlinear programming, calculus of variations, random-process theory, special functions, combinatorial analysis, game theory, much more. 1152pp. 5 3/8 x 8 1/2.   0-486-41147-8

A HEAT TRANSFER TEXTBOOK: Fourth Edition, John H. Lienhard V and John H. Lienhard IV. This introduction to heat and mass transfer for engineering students features worked examples and end-of-chapter exercises. Worked examples and end-of-chapter exercises appear throughout the book, along with well-drawn, illuminating figures. 768pp. 7 x 9 1/4.   0-486-47931-5

BASIC ELECTRICITY, U.S. Bureau of Naval Personnel. Originally a training course; best nontechnical coverage. Topics include batteries, circuits, conductors, AC and DC, inductance and capacitance, generators, motors, transformers, amplifiers, etc. Many questions with answers. 349 illustrations. 1969 edition. 448pp. 6 1/2 x 9 1/4.
0-486-20973-3

BASIC ELECTRONICS, U.S. Bureau of Naval Personnel. Clear, well-illustrated introduction to electronic equipment covers numerous essential topics: electron tubes, semiconductors, electronic power supplies, tuned circuits, amplifiers, receivers, ranging and navigation systems, computers, antennas, more. 560 illustrations. 567pp. 6 1/2 x 9 1/4.   0-486-21076-6

BASIC WING AND AIRFOIL THEORY, Alan Pope. This self-contained treatment by a pioneer in the study of wind effects covers flow functions, airfoil construction and pressure distribution, finite and monoplane wings, and many other subjects. 1951 edition. 320pp. 5 3/8 x 8 1/2.   0-486-47188-8

SYNTHETIC FUELS, Ronald F. Probstein and R. Edwin Hicks. This unified presentation examines the methods and processes for converting coal, oil, shale, tar sands, and various forms of biomass into liquid, gaseous, and clean solid fuels. 1982 edition. 512pp. 6 1/8 x 9 1/4.   0-486-44977-7

THEORY OF ELASTIC STABILITY, Stephen P. Timoshenko and James M. Gere. Written by world-renowned authorities on mechanics, this classic ranges from theoretical explanations of 2- and 3-D stress and strain to practical applications such as torsion, bending, and thermal stress. 1961 edition. 560pp. 5 3/8 x 8 1/2.   0-486-47207-8

PRINCIPLES OF DIGITAL COMMUNICATION AND CODING, Andrew J. Viterbi and Jim K. Omura. This classic by two digital communications experts is geared toward students of communications theory and to designers of channels, links, terminals, modems, or networks used to transmit and receive digital messages. 1979 edition. 576pp. 6 1/8 x 9 1/4.   0-486-46901-8

LINEAR SYSTEM THEORY: The State Space Approach, Lotfi A. Zadeh and Charles A. Desoer. Written by two pioneers in the field, this exploration of the state space approach focuses on problems of stability and control, plus connections between this approach and classical techniques. 1963 edition. 656pp. 6 1/8 x 9 1/4.
0-486-46663-9

Browse over 9,000 books at www.doverpublications.com

Mathematics-Bestsellers

HANDBOOK OF MATHEMATICAL FUNCTIONS: with Formulas, Graphs, and Mathematical Tables, Edited by Milton Abramowitz and Irene A. Stegun. A classic resource for working with special functions, standard trig, and exponential logarithmic definitions and extensions, it features 29 sets of tables, some to as high as 20 places. 1046pp. 8 x 10 1/2.   0-486-61272-4

ABSTRACT AND CONCRETE CATEGORIES: The Joy of Cats, Jiri Adamek, Horst Herrlich, and George E. Strecker. This up-to-date introductory treatment employs category theory to explore the theory of structures. Its unique approach stresses concrete categories and presents a systematic view of factorization structures. Numerous examples. 1990 edition, updated 2004. 528pp. 6 1/8 x 9 1/4.   0-486-46934-4

MATHEMATICS: Its Content, Methods and Meaning, A. D. Aleksandrov, A. N. Kolmogorov, and M. A. Lavrent'ev. Major survey offers comprehensive, coherent discussions of analytic geometry, algebra, differential equations, calculus of variations, functions of a complex variable, prime numbers, linear and non-Euclidean geometry, topology, functional analysis, more. 1963 edition. 1120pp. 5 3/8 x 8 1/2.   0-486-40916-3

INTRODUCTION TO VECTORS AND TENSORS: Second Edition--Two Volumes Bound as One, Ray M. Bowen and C.-C. Wang. Convenient single-volume compilation of two texts offers both introduction and in-depth survey. Geared toward engineering and science students rather than mathematicians, it focuses on physics and engineering applications. 1976 edition. 560pp. 6 1/2 x 9 1/4.   0-486-46914-X

AN INTRODUCTION TO ORTHOGONAL POLYNOMIALS, Theodore S. Chihara. Concise introduction covers general elementary theory, including the representation theorem and distribution functions, continued fractions and chain sequences, the recurrence formula, special functions, and some specific systems. 1978 edition. 272pp. 5 3/8 x 8 1/2.
0-486-47929-3

ADVANCED MATHEMATICS FOR ENGINEERS AND SCIENTISTS, Paul DuChateau. This primary text and supplemental reference focuses on linear algebra, calculus, and ordinary differential equations. Additional topics include partial differential equations and approximation methods. Includes solved problems. 1992 edition. 400pp. 7 1/2 x 9 1/4.   0-486-47930-7

PARTIAL DIFFERENTIAL EQUATIONS FOR SCIENTISTS AND ENGINEERS, Stanley J. Farlow. Practical text shows how to formulate and solve partial differential equations. Coverage of diffusion-type problems, hyperbolic-type problems, elliptic-type problems, numerical and approximate methods. Solution guide available upon request. 1982 edition. 414pp. 6 1/8 x 9 1/4.   0-486-67620-X

VARIATIONAL PRINCIPLES AND FREE-BOUNDARY PROBLEMS, Avner Friedman. Advanced graduate-level text examines variational methods in partial differential equations and illustrates their applications to free-boundary problems. Features detailed statements of standard theory of elliptic and parabolic operators. 1982 edition. 720pp. 6 1/8 x 9 1/4.   0-486-47853-X

LINEAR ANALYSIS AND REPRESENTATION THEORY, Steven A. Gaal. Unified treatment covers topics from the theory of operators and operator algebras on Hilbert spaces; integration and representation theory for topological groups; and the theory of Lie algebras, Lie groups, and transform groups. 1973 edition. 704pp. 6 1/8 x 9 1/4.
0-486-47851-3

CATALOG OF DOVER BOOKS

A SURVEY OF INDUSTRIAL MATHEMATICS, Charles R. MacCluer. Students learn how to solve problems they'll encounter in their professional lives with this concise single-volume treatment. It employs MATLAB and other strategies to explore typical industrial problems. 2000 edition. 384pp. 5 3/8 x 8 1/2. 0-486-47702-9

NUMBER SYSTEMS AND THE FOUNDATIONS OF ANALYSIS, Elliott Mendelson. Geared toward undergraduate and beginning graduate students, this study explores natural numbers, integers, rational numbers, real numbers, and complex numbers. Numerous exercises and appendixes supplement the text. 1973 edition. 368pp. 5 3/8 x 8 1/2. 0-486-45792-3

A FIRST LOOK AT NUMERICAL FUNCTIONAL ANALYSIS, W. W. Sawyer. Text by renowned educator shows how problems in numerical analysis lead to concepts of functional analysis. Topics include Banach and Hilbert spaces, contraction mappings, convergence, differentiation and integration, and Euclidean space. 1978 edition. 208pp. 5 3/8 x 8 1/2. 0-486-47882-3

FRACTALS, CHAOS, POWER LAWS: Minutes from an Infinite Paradise, Manfred Schroeder. A fascinating exploration of the connections between chaos theory, physics, biology, and mathematics, this book abounds in award-winning computer graphics, optical illusions, and games that clarify memorable insights into self-similarity. 1992 edition. 448pp. 6 1/8 x 9 1/4. 0-486-47204-3

SET THEORY AND THE CONTINUUM PROBLEM, Raymond M. Smullyan and Melvin Fitting. A lucid, elegant, and complete survey of set theory, this three part treatment explores axiomatic set theory, the consistency of the continuum hypothesis, and forcing and independence results. 1996 edition. 336pp. 6 x 9. 0-486-47484-4

DYNAMICAL SYSTEMS, Shlomo Sternberg. A pioneer in the field of dynamical systems discusses one-dimensional dynamics, differential equations, random walks, iterated function systems, symbolic dynamics, and Markov chains. Supplementary materials include PowerPoint slides and MATLAB exercises. 2010 edition. 272pp. 6 1/8 x 9 1/4. 0-486-47705-3

ORDINARY DIFFERENTIAL EQUATIONS, Morris Tenenbaum and Harry Pollard. Skillfully organized introductory text examines origin of differential equations, then defines basic terms and outlines general solution of a differential equation. Explores integrating factors; dilution and accretion problems; Laplace Transforms; Newton's Interpolation Formulas, more. 818pp. 5 3/8 x 8 1/2. 0-486-64940-7

MATROID THEORY, D. J. A. Welsh. Text by a noted expert describes standard examples and investigation results, using elementary proofs to develop basic matroid properties before advancing to a more sophisticated treatment. Includes numerous exercises. 1976 edition. 448pp. 5 3/8 x 8 1/2. 0-486-47439-9

THE CONCEPT OF A RIEMANN SURFACE, Hermann Weyl. This classic on the general history of functions combines function theory and geometry, forming the basis of the modern approach to analysis, geometry, and topology. 1955 edition. 208pp. 5 3/8 x 8 1/2. 0-486-47004-0

THE LAPLACE TRANSFORM, David Vernon Widder. This volume focuses on the Laplace and Stieltjes transforms, offering a highly theoretical treatment. Topics include fundamental formulas, the moment problem, monotonic functions, and Tauberian theorems. 1941 edition. 416pp. 5 3/8 x 8 1/2. 0-486-47755-X

Browse over 9,000 books at www.doverpublications.com

Mathematics–Logic and Problem Solving

PERPLEXING PUZZLES AND TANTALIZING TEASERS, Martin Gardner. Ninety-three riddles, mazes, illusions, tricky questions, word and picture puzzles, and other challenges offer hours of entertainment for youngsters. Filled with rib-tickling drawings. Solutions. 224pp. 5 3/8 x 8 1/2. 0-486-25637-5

MY BEST MATHEMATICAL AND LOGIC PUZZLES, Martin Gardner. The noted expert selects 70 of his favorite "short" puzzles. Includes The Returning Explorer, The Mutilated Chessboard, Scrambled Box Tops, and dozens more. Complete solutions included. 96pp. 5 3/8 x 8 1/2. 0-486-28152-3

THE LADY OR THE TIGER?: and Other Logic Puzzles, Raymond M. Smullyan. Created by a renowned puzzle master, these whimsically themed challenges involve paradoxes about probability, time, and change; metapuzzles; and self-referentiality. Nineteen chapters advance in difficulty from relatively simple to highly complex. 1982 edition. 240pp. 5 3/8 x 8 1/2. 0-486-47027-X

SATAN, CANTOR AND INFINITY: Mind-Boggling Puzzles, Raymond M. Smullyan. A renowned mathematician tells stories of knights and knaves in an entertaining look at the logical precepts behind infinity, probability, time, and change. Requires a strong background in mathematics. Complete solutions. 288pp. 5 3/8 x 8 1/2.
0-486-47036-9

THE RED BOOK OF MATHEMATICAL PROBLEMS, Kenneth S. Williams and Kenneth Hardy. Handy compilation of 100 practice problems, hints and solutions indispensable for students preparing for the William Lowell Putnam and other mathematical competitions. Preface to the First Edition. Sources. 1988 edition. 192pp. 5 3/8 x 8 1/2. 0-486-69415-1

KING ARTHUR IN SEARCH OF HIS DOG AND OTHER CURIOUS PUZZLES, Raymond M. Smullyan. This fanciful, original collection for readers of all ages features arithmetic puzzles, logic problems related to crime detection, and logic and arithmetic puzzles involving King Arthur and his Dogs of the Round Table. 160pp. 5 3/8 x 8 1/2.
0-486-47435-6

UNDECIDABLE THEORIES: Studies in Logic and the Foundation of Mathematics, Alfred Tarski in collaboration with Andrzej Mostowski and Raphael M. Robinson. This well-known book by the famed logician consists of three treatises: "A General Method in Proofs of Undecidability," "Undecidability and Essential Undecidability in Mathematics," and "Undecidability of the Elementary Theory of Groups." 1953 edition. 112pp. 5 3/8 x 8 1/2. 0-486-47703-7

LOGIC FOR MATHEMATICIANS, J. Barkley Rosser. Examination of essential topics and theorems assumes no background in logic. "Undoubtedly a major addition to the literature of mathematical logic." – Bulletin of the American Mathematical Society. 1978 edition. 592pp. 6 1/8 x 9 1/4. 0-486-46898-4

INTRODUCTION TO PROOF IN ABSTRACT MATHEMATICS, Andrew Wohlgemuth. This undergraduate text teaches students what constitutes an acceptable proof, and it develops their ability to do proofs of routine problems as well as those requiring creative insights. 1990 edition. 384pp. 6 1/2 x 9 1/4. 0-486-47854-8

FIRST COURSE IN MATHEMATICAL LOGIC, Patrick Suppes and Shirley Hill. Rigorous introduction is simple enough in presentation and context for wide range of students. Symbolizing sentences; logical inference; truth and validity; truth tables; terms, predicates, universal quantifiers; universal specification and laws of identity; more. 288pp. 5 3/8 x 8 1/2. 0-486-42259-3

Mathematics–Algebra and Calculus

VECTOR CALCULUS, Peter Baxandall and Hans Liebeck. This introductory text offers a rigorous, comprehensive treatment. Classical theorems of vector calculus are amply illustrated with figures, worked examples, physical applications, and exercises with hints and answers. 1986 edition. 560pp. 5 3/8 x 8 1/2. 0-486-46620-5

ADVANCED CALCULUS: An Introduction to Classical Analysis, Louis Brand. A course in analysis that focuses on the functions of a real variable, this text introduces the basic concepts in their simplest setting and illustrates its teachings with numerous examples, theorems, and proofs. 1955 edition. 592pp. 5 3/8 x 8 1/2. 0-486-44548-8

ADVANCED CALCULUS, Avner Friedman. Intended for students who have already completed a one-year course in elementary calculus, this two-part treatment advances from functions of one variable to those of several variables. Solutions. 1971 edition. 432pp. 5 3/8 x 8 1/2. 0-486-45795-8

METHODS OF MATHEMATICS APPLIED TO CALCULUS, PROBABILITY, AND STATISTICS, Richard W. Hamming. This 4-part treatment begins with algebra and analytic geometry and proceeds to an exploration of the calculus of algebraic functions and transcendental functions and applications. 1985 edition. Includes 310 figures and 18 tables. 880pp. 6 1/2 x 9 1/4. 0-486-43945-3

BASIC ALGEBRA I: Second Edition, Nathan Jacobson. A classic text and standard reference for a generation, this volume covers all undergraduate algebra topics, including groups, rings, modules, Galois theory, polynomials, linear algebra, and associative algebra. 1985 edition. 528pp. 6 1/8 x 9 1/4. 0-486-47189-6

BASIC ALGEBRA II: Second Edition, Nathan Jacobson. This classic text and standard reference comprises all subjects of a first-year graduate-level course, including in-depth coverage of groups and polynomials and extensive use of categories and functors. 1989 edition. 704pp. 6 1/8 x 9 1/4. 0-486-47187-X

CALCULUS: An Intuitive and Physical Approach (Second Edition), Morris Kline. Application-oriented introduction relates the subject as closely as possible to science with explorations of the derivative; differentiation and integration of the powers of x; theorems on differentiation, antidifferentiation; the chain rule; trigonometric functions; more. Examples. 1967 edition. 960pp. 6 1/2 x 9 1/4. 0-486-40453-6

ABSTRACT ALGEBRA AND SOLUTION BY RADICALS, John E. Maxfield and Margaret W. Maxfield. Accessible advanced undergraduate-level text starts with groups, rings, fields, and polynomials and advances to Galois theory, radicals and roots of unity, and solution by radicals. Numerous examples, illustrations, exercises, appendixes. 1971 edition. 224pp. 6 1/8 x 9 1/4. 0-486-47723-1

AN INTRODUCTION TO THE THEORY OF LINEAR SPACES, Georgi E. Shilov. Translated by Richard A. Silverman. Introductory treatment offers a clear exposition of algebra, geometry, and analysis as parts of an integrated whole rather than separate subjects. Numerous examples illustrate many different fields, and problems include hints or answers. 1961 edition. 320pp. 5 3/8 x 8 1/2. 0-486-63070-6

LINEAR ALGEBRA, Georgi E. Shilov. Covers determinants, linear spaces, systems of linear equations, linear functions of a vector argument, coordinate transformations, the canonical form of the matrix of a linear operator, bilinear and quadratic forms, and more. 387pp. 5 3/8 x 8 1/2. 0-486-63518-X

Browse over 9,000 books at www.doverpublications.com

Mathematics–Probability and Statistics

BASIC PROBABILITY THEORY, Robert B. Ash. This text emphasizes the probabilistic way of thinking, rather than measure-theoretic concepts. Geared toward advanced undergraduates and graduate students, it features solutions to some of the problems. 1970 edition. 352pp. 5 3/8 x 8 1/2. 0-486-46628-0

PRINCIPLES OF STATISTICS, M. G. Bulmer. Concise description of classical statistics, from basic dice probabilities to modern regression analysis. Equal stress on theory and applications. Moderate difficulty; only basic calculus required. Includes problems with answers. 252pp. 5 5/8 x 8 1/4. 0-486-63760-3

OUTLINE OF BASIC STATISTICS: Dictionary and Formulas, John E. Freund and Frank J. Williams. Handy guide includes a 70-page outline of essential statistical formulas covering grouped and ungrouped data, finite populations, probability, and more, plus over 1,000 clear, concise definitions of statistical terms. 1966 edition. 208pp. 5 3/8 x 8 1/2. 0-486-47769-X

GOOD THINKING: The Foundations of Probability and Its Applications, Irving J. Good. This in-depth treatment of probability theory by a famous British statistician explores Keynesian principles and surveys such topics as Bayesian rationality, corroboration, hypothesis testing, and mathematical tools for induction and simplicity. 1983 edition. 352pp. 5 3/8 x 8 1/2. 0-486-47438-0

INTRODUCTION TO PROBABILITY THEORY WITH CONTEMPORARY APPLICATIONS, Lester L. Helms. Extensive discussions and clear examples, written in plain language, expose students to the rules and methods of probability. Exercises foster problem-solving skills, and all problems feature step-by-step solutions. 1997 edition. 368pp. 6 1/2 x 9 1/4. 0-486-47418-6

CHANCE, LUCK, AND STATISTICS, Horace C. Levinson. In simple, non-technical language, this volume explores the fundamentals governing chance and applies them to sports, government, and business. "Clear and lively ... remarkably accurate." – *Scientific Monthly*. 384pp. 5 3/8 x 8 1/2. 0-486-41997-5

FIFTY CHALLENGING PROBLEMS IN PROBABILITY WITH SOLUTIONS, Frederick Mosteller. Remarkable puzzlers, graded in difficulty, illustrate elementary and advanced aspects of probability. These problems were selected for originality, general interest, or because they demonstrate valuable techniques. Also includes detailed solutions. 88pp. 5 3/8 x 8 1/2. 0-486-65355-2

EXPERIMENTAL STATISTICS, Mary Gibbons Natrella. A handbook for those seeking engineering information and quantitative data for designing, developing, constructing, and testing equipment. Covers the planning of experiments, the analyzing of extreme-value data; and more. 1966 edition. Index. Includes 52 figures and 76 tables. 560pp. 8 3/8 x 11. 0-486-43937-2

STOCHASTIC MODELING: Analysis and Simulation, Barry L. Nelson. Coherent introduction to techniques also offers a guide to the mathematical, numerical, and simulation tools of systems analysis. Includes formulation of models, analysis, and interpretation of results. 1995 edition. 336pp. 6 1/8 x 9 1/4. 0-486-47770-3

INTRODUCTION TO BIOSTATISTICS: Second Edition, Robert R. Sokal and F. James Rohlf. Suitable for undergraduates with a minimal background in mathematics, this introduction ranges from descriptive statistics to fundamental distributions and the testing of hypotheses. Includes numerous worked-out problems and examples. 1987 edition. 384pp. 6 1/8 x 9 1/4. 0-486-46961-1

Browse over 9,000 books at www.doverpublications.com

Mathematics–Geometry and Topology

PROBLEMS AND SOLUTIONS IN EUCLIDEAN GEOMETRY, M. N. Aref and William Wernick. Based on classical principles, this book is intended for a second course in Euclidean geometry and can be used as a refresher. More than 200 problems include hints and solutions. 1968 edition. 272pp. 5 3/8 x 8 1/2.   0-486-47720-7

TOPOLOGY OF 3-MANIFOLDS AND RELATED TOPICS, Edited by M. K. Fort, Jr. With a New Introduction by Daniel Silver. Summaries and full reports from a 1961 conference discuss decompositions and subsets of 3-space; n-manifolds; knot theory; the Poincaré conjecture; and periodic maps and isotopies. Familiarity with algebraic topology required. 1962 edition. 272pp. 6 1/8 x 9 1/4.   0-486-47753-3

POINT SET TOPOLOGY, Steven A. Gaal. Suitable for a complete course in topology, this text also functions as a self-contained treatment for independent study. Additional enrichment materials make it equally valuable as a reference. 1964 edition. 336pp. 5 3/8 x 8 1/2.   0-486-47222-1

INVITATION TO GEOMETRY, Z. A. Melzak. Intended for students of many different backgrounds with only a modest knowledge of mathematics, this text features self-contained chapters that can be adapted to several types of geometry courses. 1983 edition. 240pp. 5 3/8 x 8 1/2.   0-486-46626-4

TOPOLOGY AND GEOMETRY FOR PHYSICISTS, Charles Nash and Siddhartha Sen. Written by physicists for physics students, this text assumes no detailed background in topology or geometry. Topics include differential forms, homotopy, homology, cohomology, fiber bundles, connection and covariant derivatives, and Morse theory. 1983 edition. 320pp. 5 3/8 x 8 1/2.   0-486-47852-1

BEYOND GEOMETRY: Classic Papers from Riemann to Einstein, Edited with an Introduction and Notes by Peter Pesic. This is the only English-language collection of these 8 accessible essays. They trace seminal ideas about the foundations of geometry that led to Einstein's general theory of relativity. 224pp. 6 1/8 x 9 1/4.   0-486-45350-2

GEOMETRY FROM EUCLID TO KNOTS, Saul Stahl. This text provides a historical perspective on plane geometry and covers non-neutral Euclidean geometry, circles and regular polygons, projective geometry, symmetries, inversions, informal topology, and more. Includes 1,000 practice problems. Solutions available. 2003 edition. 480pp. 6 1/8 x 9 1/4.   0-486-47459-3

TOPOLOGICAL VECTOR SPACES, DISTRIBUTIONS AND KERNELS, François Trèves. Extending beyond the boundaries of Hilbert and Banach space theory, this text focuses on key aspects of functional analysis, particularly in regard to solving partial differential equations. 1967 edition. 592pp. 5 3/8 x 8 1/2.
0-486-45352-9

INTRODUCTION TO PROJECTIVE GEOMETRY, C. R. Wylie, Jr. This introductory volume offers strong reinforcement for its teachings, with detailed examples and numerous theorems, proofs, and exercises, plus complete answers to all odd-numbered end-of-chapter problems. 1970 edition. 576pp. 6 1/8 x 9 1/4.   0-486-46895-X

FOUNDATIONS OF GEOMETRY, C. R. Wylie, Jr. Geared toward students preparing to teach high school mathematics, this text explores the principles of Euclidean and non-Euclidean geometry and covers both generalities and specifics of the axiomatic method. 1964 edition. 352pp. 6 x 9.   0-486-47214-0

Browse over 9,000 books at www.doverpublications.com

Mathematics-History

THE WORKS OF ARCHIMEDES, Archimedes. Translated by Sir Thomas Heath. Complete works of ancient geometer feature such topics as the famous problems of the ratio of the areas of a cylinder and an inscribed sphere; the properties of conoids, spheroids, and spirals; more. 326pp. 5 3/8 x 8 1/2. 0-486-42084-1

THE HISTORICAL ROOTS OF ELEMENTARY MATHEMATICS, Lucas N. H. Bunt, Phillip S. Jones, and Jack D. Bedient. Exciting, hands-on approach to understanding fundamental underpinnings of modern arithmetic, algebra, geometry and number systems examines their origins in early Egyptian, Babylonian, and Greek sources. 336pp. 5 3/8 x 8 1/2. 0-486-25563-8

THE THIRTEEN BOOKS OF EUCLID'S ELEMENTS, Euclid. Contains complete English text of all 13 books of the Elements plus critical apparatus analyzing each definition, postulate, and proposition in great detail. Covers textual and linguistic matters; mathematical analyses of Euclid's ideas; classical, medieval, Renaissance and modern commentators; refutations, supports, extrapolations, reinterpretations and historical notes. 995 figures. Total of 1,425pp. All books 5 3/8 x 8 1/2.

Vol. I: 443pp. 0-486-60088-2
Vol. II: 464pp. 0-486-60089-0
Vol. III: 546pp. 0-486-60090-4

A HISTORY OF GREEK MATHEMATICS, Sir Thomas Heath. This authoritative two-volume set that covers the essentials of mathematics and features every landmark innovation and every important figure, including Euclid, Apollonius, and others. 5 3/8 x 8 1/2.

Vol. I: 461pp. 0-486-24073-8
Vol. II: 597pp. 0-486-24074-6

A MANUAL OF GREEK MATHEMATICS, Sir Thomas L. Heath. This concise but thorough history encompasses the enduring contributions of the ancient Greek mathematicians whose works form the basis of most modern mathematics. Discusses Pythagorean arithmetic, Plato, Euclid, more. 1931 edition. 576pp. 5 3/8 x 8 1/2.

0-486-43231-9

CHINESE MATHEMATICS IN THE THIRTEENTH CENTURY, Ulrich Libbrecht. An exploration of the 13th-century mathematician Ch'in, this fascinating book combines what is known of the mathematician's life with a history of his only extant work, the Shu-shu chiu-chang. 1973 edition. 592pp. 5 3/8 x 8 1/2.

0-486-44619-0

PHILOSOPHY OF MATHEMATICS AND DEDUCTIVE STRUCTURE IN EUCLID'S ELEMENTS, Ian Mueller. This text provides an understanding of the classical Greek conception of mathematics as expressed in Euclid's Elements. It focuses on philosophical, foundational, and logical questions and features helpful appendixes. 400pp. 6 1/2 x 9 1/4. 0-486-45300-6

BEYOND GEOMETRY: Classic Papers from Riemann to Einstein, Edited with an Introduction and Notes by Peter Pesic. This is the only English-language collection of these 8 accessible essays. They trace seminal ideas about the foundations of geometry that led to Einstein's general theory of relativity. 224pp. 6 1/8 x 9 1/4. 0-486-45350-2

HISTORY OF MATHEMATICS, David E. Smith. Two-volume history – from Egyptian papyri and medieval maps to modern graphs and diagrams. Non-technical chronological survey with thousands of biographical notes, critical evaluations, and contemporary opinions on over 1,100 mathematicians. 5 3/8 x 8 1/2.

Vol. I: 618pp. 0-486-20429-4
Vol. II: 736pp. 0-486-20430-8

Physics

THEORETICAL NUCLEAR PHYSICS, John M. Blatt and Victor F. Weisskopf. An uncommonly clear and cogent investigation and correlation of key aspects of theoretical nuclear physics by leading experts: the nucleus, nuclear forces, nuclear spectroscopy, two-, three- and four-body problems, nuclear reactions, beta-decay and nuclear shell structure. 896pp. 5 3/8 x 8 1/2. 0-486-66827-4

QUANTUM THEORY, David Bohm. This advanced undergraduate-level text presents the quantum theory in terms of qualitative and imaginative concepts, followed by specific applications worked out in mathematical detail. 655pp. 5 3/8 x 8 1/2.
0-486-65969-0

ATOMIC PHYSICS AND HUMAN KNOWLEDGE, Niels Bohr. Articles and speeches by the Nobel Prize–winning physicist, dating from 1934 to 1958, offer philosophical explorations of the relevance of atomic physics to many areas of human endeavor. 1961 edition. 112pp. 5 3/8 x 8 1/2. 0-486-47928-5

COSMOLOGY, Hermann Bondi. A co-developer of the steady-state theory explores his conception of the expanding universe. This historic book was among the first to present cosmology as a separate branch of physics. 1961 edition. 192pp. 5 3/8 x 8 1/2.
0-486-47483-6

LECTURES ON QUANTUM MECHANICS, Paul A. M. Dirac. Four concise, brilliant lectures on mathematical methods in quantum mechanics from Nobel Prize-winning quantum pioneer build on idea of visualizing quantum theory through the use of classical mechanics. 96pp. 5 3/8 x 8 1/2. 0-486-41713-1

THE PRINCIPLE OF RELATIVITY, Albert Einstein and Frances A. Davis. Eleven papers that forged the general and special theories of relativity include seven papers by Einstein, two by Lorentz, and one each by Minkowski and Weyl. 1923 edition. 240pp. 5 3/8 x 8 1/2. 0-486-60081-5

PHYSICS OF WAVES, William C. Elmore and Mark A. Heald. Ideal as a classroom text or for individual study, this unique one-volume overview of classical wave theory covers wave phenomena of acoustics, optics, electromagnetic radiations, and more. 477pp. 5 3/8 x 8 1/2. 0-486-64926-1

THERMODYNAMICS, Enrico Fermi. In this classic of modern science, the Nobel Laureate presents a clear treatment of systems, the First and Second Laws of Thermodynamics, entropy, thermodynamic potentials, and much more. Calculus required. 160pp. 5 3/8 x 8 1/2. 0-486-60361-X

QUANTUM THEORY OF MANY-PARTICLE SYSTEMS, Alexander L. Fetter and John Dirk Walecka. Self-contained treatment of nonrelativistic many-particle systems discusses both formalism and applications in terms of ground-state (zero-temperature) formalism, finite-temperature formalism, canonical transformations, and applications to physical systems. 1971 edition. 640pp. 5 3/8 x 8 1/2. 0-486-42827-3

QUANTUM MECHANICS AND PATH INTEGRALS: Emended Edition, Richard P. Feynman and Albert R. Hibbs. Emended by Daniel F. Styer. The Nobel Prize–winning physicist presents unique insights into his theory and its applications. Feynman starts with fundamentals and advances to the perturbation method, quantum electrodynamics, and statistical mechanics. 1965 edition, emended in 2005. 384pp. 6 1/8 x 9 1/4. 0-486-47722-3

Physics

INTRODUCTION TO MODERN OPTICS, Grant R. Fowles. A complete basic undergraduate course in modern optics for students in physics, technology, and engineering. The first half deals with classical physical optics; the second, quantum nature of light. Solutions. 336pp. 5 3/8 x 8 1/2.   0-486-65957-7

THE QUANTUM THEORY OF RADIATION: Third Edition, W. Heitler. The first comprehensive treatment of quantum physics in any language, this classic introduction to basic theory remains highly recommended and widely used, both as a text and as a reference. 1954 edition. 464pp. 5 3/8 x 8 1/2.   0-486-64558-4

QUANTUM FIELD THEORY, Claude Itzykson and Jean-Bernard Zuber. This comprehensive text begins with the standard quantization of electrodynamics and perturbative renormalization, advancing to functional methods, relativistic bound states, broken symmetries, nonabelian gauge fields, and asymptotic behavior. 1980 edition. 752pp. 6 1/2 x 9 1/4.   0-486-44568-2

FOUNDATIONS OF POTENTIAL THERY, Oliver D. Kellogg. Introduction to fundamentals of potential functions covers the force of gravity, fields of force, potentials, harmonic functions, electric images and Green's function, sequences of harmonic functions, fundamental existence theorems, and much more. 400pp. 5 3/8 x 8 1/2.
                                                                  0-486-60144-7

FUNDAMENTALS OF MATHEMATICAL PHYSICS, Edgar A. Kraut. Indispensable for students of modern physics, this text provides the necessary background in mathematics to study the concepts of electromagnetic theory and quantum mechanics. 1967 edition. 480pp. 6 1/2 x 9 1/4.   0-486-45809-1

GEOMETRY AND LIGHT: The Science of Invisibility, Ulf Leonhardt and Thomas Philbin. Suitable for advanced undergraduate and graduate students of engineering, physics, and mathematics and scientific researchers of all types, this is the first authoritative text on invisibility and the science behind it. More than 100 full-color illustrations, plus exercises with solutions. 2010 edition. 288pp. 7 x 9 1/4.   0-486-47693-6

QUANTUM MECHANICS: New Approaches to Selected Topics, Harry J. Lipkin. Acclaimed as "excellent" (*Nature*) and "very original and refreshing" (*Physics Today*), these studies examine the Mössbauer effect, many-body quantum mechanics, scattering theory, Feynman diagrams, and relativistic quantum mechanics. 1973 edition. 480pp. 5 3/8 x 8 1/2.   0-486-45893-8

THEORY OF HEAT, James Clerk Maxwell. This classic sets forth the fundamentals of thermodynamics and kinetic theory simply enough to be understood by beginners, yet with enough subtlety to appeal to more advanced readers, too. 352pp. 5 3/8 x 8 1/2.   0-486-41735-2

QUANTUM MECHANICS, Albert Messiah. Subjects include formalism and its interpretation, analysis of simple systems, symmetries and invariance, methods of approximation, elements of relativistic quantum mechanics, much more. "Strongly recommended." – *American Journal of Physics.* 1152pp. 5 3/8 x 8 1/2.   0-486-40924-4

RELATIVISTIC QUANTUM FIELDS, Charles Nash. This graduate-level text contains techniques for performing calculations in quantum field theory. It focuses chiefly on the dimensional method and the renormalization group methods. Additional topics include functional integration and differentiation. 1978 edition. 240pp. 5 3/8 x 8 1/2.
                                                                  0-486-47752-5

Physics

MATHEMATICAL TOOLS FOR PHYSICS, James Nearing. Encouraging students' development of intuition, this original work begins with a review of basic mathematics and advances to infinite series, complex algebra, differential equations, Fourier series, and more. 2010 edition. 496pp. 6 1/8 x 9 1/4. 0-486-48212-X

TREATISE ON THERMODYNAMICS, Max Planck. Great classic, still one of the best introductions to thermodynamics. Fundamentals, first and second principles of thermodynamics, applications to special states of equilibrium, more. Numerous worked examples. 1917 edition. 297pp. 5 3/8 x 8. 0-486-66371-X

AN INTRODUCTION TO RELATIVISTIC QUANTUM FIELD THEORY, Silvan S. Schweber. Complete, systematic, and self-contained, this text introduces modern quantum field theory. "Combines thorough knowledge with a high degree of didactic ability and a delightful style." – *Mathematical Reviews*. 1961 edition. 928pp. 5 3/8 x 8 1/2. 0-486-44228-4

THE ELECTROMAGNETIC FIELD, Albert Shadowitz. Comprehensive undergraduate text covers basics of electric and magnetic fields, building up to electromagnetic theory. Related topics include relativity theory. Over 900 problems, some with solutions. 1975 edition. 768pp. 5 5/8 x 8 1/4. 0-486-65660-8

THE PRINCIPLES OF STATISTICAL MECHANICS, Richard C. Tolman. Definitive treatise offers a concise exposition of classical statistical mechanics and a thorough elucidation of quantum statistical mechanics, plus applications of statistical mechanics to thermodynamic behavior. 1930 edition. 704pp. 5 5/8 x 8 1/4. 0-486-63896-0

INTRODUCTION TO THE PHYSICS OF FLUIDS AND SOLIDS, James S. Trefil. This interesting, informative survey by a well-known science author ranges from classical physics and geophysical topics, from the rings of Saturn and the rotation of the galaxy to underground nuclear tests. 1975 edition. 320pp. 5 3/8 x 8 1/2. 0-486-47437-2

STATISTICAL PHYSICS, Gregory H. Wannier. Classic text combines thermodynamics, statistical mechanics, and kinetic theory in one unified presentation. Topics include equilibrium statistics of special systems, kinetic theory, transport coefficients, and fluctuations. Problems with solutions. 1966 edition. 532pp. 5 3/8 x 8 1/2. 0-486-65401-X

SPACE, TIME, MATTER, Hermann Weyl. Excellent introduction probes deeply into Euclidean space, Riemann's space, Einstein's general relativity, gravitational waves and energy, and laws of conservation. "A classic of physics." – *British Journal for Philosophy and Science*. 330pp. 5 3/8 x 8 1/2. 0-486-60267-2

RANDOM VIBRATIONS: Theory and Practice, Paul H. Wirsching, Thomas L. Paez and Keith Ortiz. Comprehensive text and reference covers topics in probability, statistics, and random processes, plus methods for analyzing and controlling random vibrations. Suitable for graduate students and mechanical, structural, and aerospace engineers. 1995 edition. 464pp. 5 3/8 x 8 1/2. 0-486-45015-5

PHYSICS OF SHOCK WAVES AND HIGH-TEMPERATURE HYDRO DYNAMIC PHENOMENA, Ya B. Zel'dovich and Yu P. Raizer. Physical, chemical processes in gases at high temperatures are focus of outstanding text, which combines material from gas dynamics, shock-wave theory, thermodynamics and statistical physics, other fields. 284 illustrations. 1966–1967 edition. 944pp. 6 1/8 x 9 1/4. 0-486-42002-7

Browse over 9,000 books at www.doverpublications.com